Naturalist on the Nanticoke

Naturalist
ON THE
Nanticoke

THE NATURAL HISTORY OF A RIVER
ON MARYLAND'S EASTERN SHORE

BY
Robert A. Hedeen

TIDEWATER PUBLISHERS

Centreville, Maryland

Copyright © 1982 by Tidewater Publishers

All rights reserved. No part of this book may be used or reproduced
in any manner whatsoever without written permission
except in the case of brief quotations embodied in critical articles and reviews
For information, address Tidewater Publishers,
Centreville, Maryland 21617.

An earlier version of Chapter 5 appeared
in the January 1981 issue of *Fur-Fish-Game* magazine.

Library of Congress Cataloging in Publication Data

Hedeen, Robert A., 1928-
Naturalist on the Nanticoke

Includes index.
1. Stream ecology—Nanticoke River (Del. and Md.)
2. Natural History—Nanticoke River (Del. and Md.)
3. Nanticoke River (Del. and Md.) I. Title.
QH105.M3H4   508.752′2   81-18524
ISBN 0-87033-282-1   AACR2

The drawings are by Julie Payne

Manufactured in the United States of America

First edition

To

Peggy

who has been most understanding

Contents

The Nanticoke (maps) : *viii-ix*

Foreword : *xi*

Introduction : 3

1 : The Alpha and Omega of It All : 9

2 : Sea Nettles and Jug-Stoppers : 17

3 : The Horseshoe Crab, A Living Fossil : 30

4 : Stilettoes, Meataxes, and Bayonets : 42

5 : Garfish Marauders : 64

6 : Poor Man's Tarpon : 78

7 : Cats, Toads, and Stingrays : 88

8 : Where Have All the Rockfish Gone? : 106

9 : Of Turtles and Snakes : 119

10 : The Return of the Erne : 136

11 : The Big Adventure; Is It Worth It? : 144

Appendix : Scientific Names of Animals : 161

Index : 165

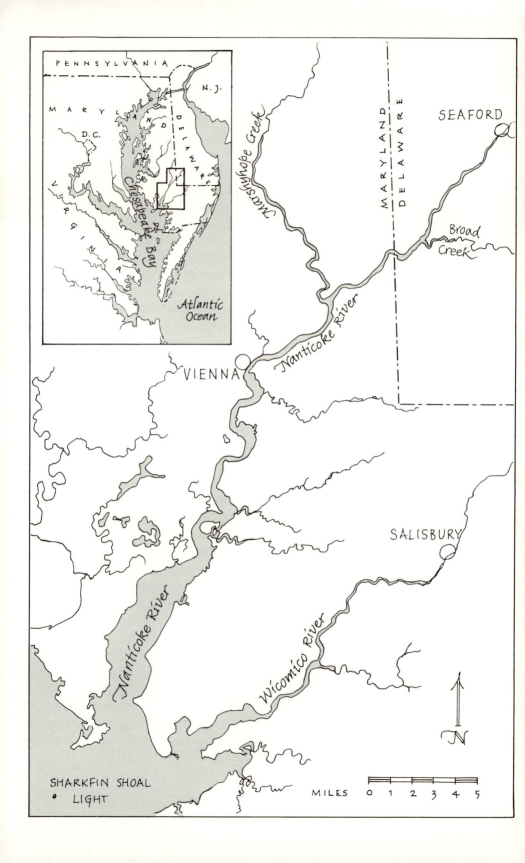

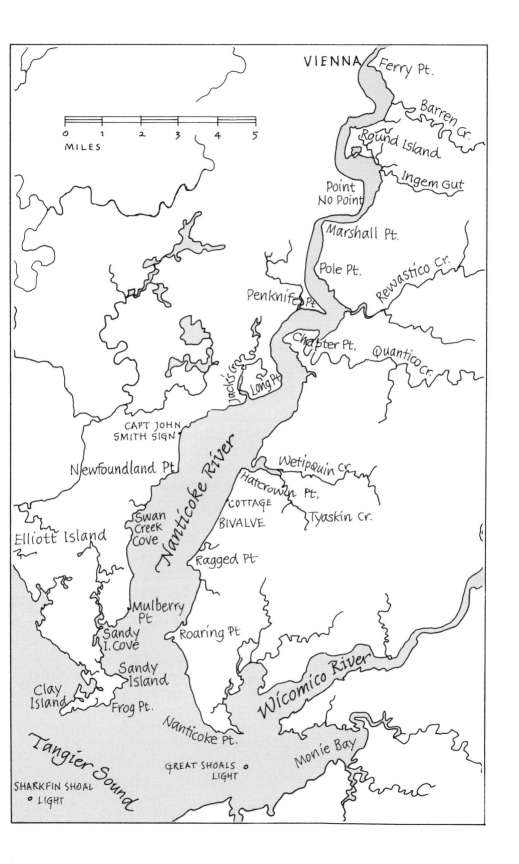

Foreword

SHORTLY AFTER I acquired possession of my tiny stretch of waterfront property between the villages of Tyaskin and Bivalve, a neighbor told me that over on the west bank of the river in Dorchester County there was some sort of historical marker that indicated the approximate landing site of Captain John Smith in June 1608. He said he had seen the marker, a small sign on a metal post, some years ago, but with the way things change on the river, he could not remember the exact spot.

Several expeditions to the west bank in search of the sign proved fruitless, but, incurable romantic that I am, I was determined to find it. Some months later while conversing with a retired waterman who frequents the dock at Bivalve Harbor, I inquired about the location of the sign. He was able to give me more explicit instructions. "Just short of Newfoundland Point and a mite north of the gut if she ain't washed into the river fore now." The next day I headed across the water in my little outboard-powered skiff and landed on a small stretch of beach at the mouth of a gut or rivulet below the point. As I beached the craft I was attacked immediately—not by savage Nantiquake Indians but by countless legions of bloodthirsty salt-marsh mosquitoes and greenhead flies. The Nantiquakes had various ways of dealing with these insects they called pootzahs (one way was to move to the windswept coast of the Atlantic Ocean during the summer months), but the only method to prevent exsanguination I had at hand was a small bottle of repellent I had stowed in the boat. I covered every inch of exposed skin with the foul-smelling material, but it was only partially effective. The chemical did, however, discourage enough of the voracious dipterans that I was able to make my way inland through a scrubby stand of cottonwoods toward the edge of the vast sea of marshland.

As luck would have it, I stumbled onto the old rusting sign in a manner of minutes. It was weather-beaten, pierced by several bullet holes, and the printing on it was badly faded from years of exposure. I managed to catch the light just right on the printing and eventually was able to decipher the message, "Captian John Smith Was Here, June 1608." I studied the writing, noting that captain was misspelled. There was no indication who had erected this historical marker so many years before. But, whoever it was would probably have received no thanks from the pompous and priggish Captain Smith. Undoubtedly, he would have been outraged at the distortion of his official title. In any event, however, this was probably the approximate spot where Smith and his companions came ashore some three hundred sixty years previously.

The diary of this colonial adventurer reveals he explored briefly this area of the Nanticoke in the late spring of 1608. With fourteen men in an open sailing pinnace, he left the settlement at Jamestown in the early part of June. Heading north in search of a passage to the Far East, gold, silver, jewels, slaves, or anything else that would bring him fame and fortune, he discovered the mouth of the Pocomoke River where it empties into Tangier Sound near the present day Maryland-Virginia boundary. Finding the Pocomoke to be surrounded by nothing but swamps and huge cypress trees, he returned to the open waters of the Sound and sailed north. A day later he arrived at the confluence of the Wicomico and Nanticoke rivers which were much larger at their mouths than the Pocomoke. For some reason he chose to head northwest up the Nanticoke and ignored the Wicomico.

He did not proceed too far up the estuary as his description of the voyage reveals an encounter with a group of Indians at a point just before the river narrows drastically. This location

"Captian John Smith Was Here June 1608"

would have to be somewhere between Newfoundland and Long points. Apparently the "savages" attempted to lure Smith to the shore but were greeted by a volley of gunshots. Sometime later Smith went ashore and records finding "much blood but not a savage." This encounter discouraged the explorers, as did the scarcity of fresh water, and so they withdrew to the south and sailed to the western shore of the Bay through what is known as Hooper's Straits.

As I stood there by the old rusted sign, fantasizing about events that had transpired so long ago, the perspiration was flowing profusely, and the repellent I had smeared on my skin was becoming diluted and less effective. The pootzahs were breaking through the chemical barrier, driving their red-hot proboscises into me in search of a blood-filled capillary. It definitely was time to retreat. Hastily I shot a photograph of the sign to preserve it on film and hurried back to the boat. Luckily the motor started on the second pull, and I sped away to the safety of the open water. A swarm of disgruntled, hungry flies followed the speeding boat for a quarter of a mile out into the river, but apparently finally became discouraged and returned to their breeding grounds in the marsh.

As far as can be determined, Captain Smith and his crew were the first Europeans to venture into the region of the Nanticoke, and I mused as I returned to the eastern shore that if that gallant captain could come back to this spot on the river today, it is quite likely he would recognize it. There has been little change in the physiography of the low lying marshes since he was there. Marshes change very little with the passage of time, and their hostile character effectively discourages the wholesale invasion and subsequent destructive practices of man. Men do enter the marshes for short periods of time to hunt, trap, and fish in the guts, but they soon depart and leave little evidence of their visits. About the only reminder of man's early presence is an occasional arrowhead one finds washed up on the beach. Incidentally, the beach in front of the Smith sign is the best place to find these artifacts anywhere on the river. Dr. Dan Sistrunk of Salisbury found an English penny on the beach in front of my cottage, and he was sure this coin bearing

Foreword

the date 1725 could be sold for a fortune. However, his hopes of early retirement were dashed when upon consulting a coin catalog he discovered his find was listed at $1.40 for a specimen in excellent condition.

Scattered duck blinds and hunting shacks give testimony to man's visits to the marshes today, but these structures are ephemeral and sooner or later disappear as the forces of nature act to restore the landscape to its primeval state. Captain John Smith would recognize the marshes, but he would be dumbfounded by the changes that have occurred on the fastland. Almost every iota of land on which a house can be built has been divided, subdivided, and subdivided again. Houses are built wherever possible and they vary from palatial domiciles to weather-beaten mobile homes. In the late 1970s lots with about one hundred feet of frontage on the water were selling for approximately ten thousand dollars, and the price will undoubtedly go higher in the future.

I feel sure much of the biologically rich marshland would have been taken years ago for construction sites if the stringent regulations against this sort of thing were not rigidly enforced by concerned governmental agencies.

Yet, in spite of the growth of the population and the extensive structural development of the fastland surrounding the river, I doubt if there are more than two or three species of animals present in the early 1600s that are not with us today. Certainly, the numbers of many species have declined—the ducks, geese, and swans no longer literally darken the sky during the fall as they did in the past—but most animal types have managed to adapt to the changing conditions, as they have for countless ages, and continue to survive.

Not all types of wildlife have decreased in numbers. Natural resources experts believe the numbers of several species have actually increased since colonial times. The white-tailed deer is a prime example of a highly adaptable species, and hunting regulations have become increasingly more lenient during the past generation. Today deer are so plentiful along the banks of the Nanticoke they are sometimes considered a threat to agricultural operations. Farmers frequently apply for and are granted special permits to kill deer out of season if they are able to demonstrate their crops are suffering as a result of browsing and feeding by the cervines.

Trappers of yore took about as many furbearers from the marshes as do those hardy souls who venture forth today into the quagmires with trap and snare in hand. Muskrat, skunk, raccoon, and fox are presently taken in goodly numbers, and many of the local watermen and farmers supplement their incomes by the sale of the pelts they extract from the land. A good day's trapping may well result in ten "rats" and a couple of coon or a fox. In 1979 the going price for a prime muskrat pelt was about eight dollars, and the stretched hide of a coon in good condition brought around fifteen. In addition, the body of the muskrat is highly prized as a food item by many residents of the Eastern Shore. It is frequently marketed under the name of swamp rabbit, and each carcass is worth about one dollar wholesale. Add up these figures and you arrive at a sum of over one hundred dollars resulting from a day's trapping, and that is not bad pay any way you look at it. With very little capital outlay and a frequent reluctance to report the exact amount received from the fur buyer, a trapper can do very well financially.

It was in this area near Tyaskin that my wife and I decided to buy our cottage in the spring of 1977. We considered ourselves fortunate in locating the ninety-four feet of waterfront with a ramshackle house attached, and we sincerely lamented with the previous owner as he enumerated the various reasons why he was compelled to dispose of the property. We had looked long and hard over the years for such a place, but, somehow or other, each location we inspected had deficiencies that were unacceptable to us. When we finally found our way to the place at Tyaskin, which the real estate agent described as a diamond in the rough, we knew our search was ended.

We each had our own reasons for wanting a bit of property like this, and these reasons were basically quite different. As for me, it was a place of my own where I could hunt, fish, boat,

Author's cottage on the Nanticoke

and, as a biologist, have a ringside seat from which I could observe and actually participate in the ongoing drama of the natural history of the lower region of the Chesapeake. As for my wife, she wanted the cottage because I wanted it so much.

After five years of hard work we have transformed the dilapidated house into a comfortable dwelling. It's amazing what some paint, discount paneling, and a fireplace can do to convert a neglected shack into a very livable cottage. At first we had no intention of using the place at any other time than in the warm months of the year, but as our fondness for the new environment changed to an infatuation, we found ourselves spending more and more time there. On one occasion we even mushed the last quarter mile through the drifted snow.

You do not just fix up a place and then have all the time in the world to sit back and relax or engage in preferred activities. Anyone who has lived on the water knows the landholder must fight a running battle with the forces of nature. The river relentlessly eats away at your property, and you must constantly engage in all-out war with it. A huge pine tree crashing into the water somewhere up on the beach during the night as it gives up the battle to the eroding and undermining forces of the tides is very disconcerting. Frequently the river is calm and peaceful in a way that deludes you into believing it will always be that way. But, you know a storm can come out of nowhere and whip the tranquil waters into waves of unbelievable, destructive power. Bulkheads, docks, and jetties take the brunt of the pounding of the waves, and when they are damaged, they must be repaired as quickly as possible or your valuable land will become just so much silt washing away, eventually out to sea.

Homeowners routinely walk the beaches, especially after a big blow, to collect bits of lumber and pilings that had once made up riverfront structures and have now been washed ashore. The materials are stacked in the yard and offer a ready supply of items needed for repair of protective waterfront structures. It is not a question of "if" your bulkhead or jetty will be damaged or torn away, it's a question only of "when."

Foreword

In spite of the necessary chores one must continually perform, there is time to enjoy the surroundings. Fortunately, our cottage was constructed with a large lower area so that the main portion of the house is well above the high-water marks made by previous floods. This extensive basement area was converted into a laboratory where I can study specimens in detail and either maintain them alive or preserve them for future examination.

Natural history is a vigorous and forceful phenomenon that one could spend every minute of his existence observing. It is more than making a checklist of birds sighted during excursions afield or making a collection of shells or arrowheads found on the beach. Natural history in its complete sense is a broad subject that encompasses the study of all living organisms in a given environment. All living things are interdependent on each other, and it is impossible to understand the processes and adaptations of one without taking into consideration the others. Living entities are woven together into that mysterious web of life we call an ecosystem, and what affects one species most assuredly has an impact on all the others.

I would like to reiterate, the primary purpose of this book is to relate my observations on only a few of the species that inhabit the region of the Nanticoke. Perhaps it will be possible to continue the story with other organisms at another time.

—Robert A. Hedeen

Tyaskin, Maryland
June, 1981

Naturalist on the Nanticoke

Introduction

THE NANTICOKE RIVER is the least well known of the major waterways that traverse the Delmarva Peninsula of the eastern United States. It originates in the heart of Sussex County in southwestern Delaware from a myriad of creeks and streams with names like Herring Run, Deep Creek, Gravelly Branch, Hurley Drain, and Gum Branch. It is navigable to barges from the city of Seaford, Delaware, for thirty-six serpentine miles as it wends its way to Tangier Sound, an offshoot of the Chesapeake Bay. Most of the river lies in the state of Maryland, and there is very little fastland abutting its banks. For the most part it is surrounded by a flat sea of marshland that seems to go on indefinitely before it finally melds with the distant skyline. The Nanticoke has always been a lonely body of water and still is.

From Tangier Sound to Long Point the river is more than a mile in width, but at this spot it suddenly narrows considerably and is no more than a quarter of a mile in width the next thirty miles to Seaford.

Captain John Smith named the river Kuskarawaok, and the Indians he encountered along its banks the Nantiquakes (Those Who Ply the Tidal Stream). Whether they deserved it or not, the Nantiquakes acquired a reputation as a fierce, warlike, and unfriendly tribe, and this, coupled with the paucity of high or fastland, discouraged settlement by the early colonials. The Indians preferred fishing and trapping to agriculture and found the lonely land well suited to their way of life.

The white man eventually moved into the area, however. In 1706 one hundred acres of land were purchased from the Indians for a reported five thousand pounds of tobacco, and Emperor's Landing was founded. Emperor's Landing was located on a broad deep stretch of the river some fourteen miles from the present-day Maryland-Delaware boundary. The name of the settlement was eventually changed to Baltimore and then to Vienna. With Vienna providing a port with an excellent route to the sea, more and more landings with accompanying settlements were established. Ellis Wharf, Walnut Landing, Wheatley's Wharf, and Lewis Landing were some of the new outposts whose names are still found on current maps and which can be visited by traveling the back roads.

The lore of the Nanticoke is scantier than that of other rivers on the peninsula, and, in fact, history does not record any significant event that transpired along its length. Blackbeard the Pirate is supposed to have frequented the area from time to time burying innumerable treasure chests, but no one has ever found one. Today some of the natives of the region still have hopes of stumbling onto a chest full of gold and silver and keep a sharp lookout for clues whenever they are in an area where a coffer is supposed to be buried. However, the fact of the matter is there is no record of the infamous Blackbeard ever venturing north of the Carolinas.

Perhaps the most infamous personality along the river during the nineteenth century was Patty Cannon. She and her new husband arrived in the area in the early 1800s and established Cannon's Ferry across the river near the Maryland-Delaware line. When she grew tired of her spouse's puritanical ways, she dispatched him with poisoned ale, according to legend, and then opened up a "low tavern" near the ferry, catering to the baser instincts of mankind. Her most notorious villainy became that of robbing and murdering slave kidnappers who frequented the area and who usually carried large sums of money. Later, when word spread that Patty Cannon's establishment should be avoided if one valued one's life, her enterprise fell off drastically. Looking for another source of

Introduction

income, she decided to set up her own slave operation. A gang of Mrs. Cannon's employees would make frequent trips to Philadelphia in search of freed Negroes. Plied with liquor laced with opiates, unsuspecting ex-slaves were smuggled back to Walnut Landing on Marshyhope Creek in Maryland, a few miles above Vienna. From there they would be transported by boat to areas in the South where they were sold illegally. Apparently this clandestine operation continued for many years and was finally terminated when Patty Cannon was arrested and charged with her crimes. In the end she confessed to having killed eleven people herself (including her husband and one of her infant children) and assisting in the murder of many others. She was sentenced to death, but before she could be executed, she cheated the hangman's noose by ingesting a lethal dose of poison. She died in agony in 1829.

Over the years other settlements were established along the estuary, mainly in Maryland. Seaford, Delaware, however, was the largest, and today it is the destination of most of the barges you see heading up the river. Vienna is the largest town along the river in Maryland, with Sharptown second in size. Sharptown and the Nanticoke River gained some notoriety a few years ago when it was discovered thousands of gallons of toxic materials known as PCBs (polychlorinated biphenyls) were illegally stored only a few feet from the river's edge in rapidly deteriorating storage tanks located within the city limits. The escape of any significant amount of the PCBs into the river would have resulted in an ecological disaster. The tides would have spread the material from one end of the river to the other, and the fin- and shellfish that were not killed outright would have been rendered unsafe for human consumption for years to come, as would have been countless generations of their offspring. Primarily due to the loud cries for action from the mayor and city council of Sharptown, television cameras and newspaper reporters from the Western Shore descended on Sharptown to report the story. Eventually the Coast Guard supervised the movement of the toxins to a more secure storage area near Baltimore. Many people in the Washington, Baltimore, and Philadelphia areas had never heard of the Nanticoke River (much less Sharptown) until the PCBs incident, and I imagine by now most of them have forgotten such places exist.

Today, as in the past, few pleasure boats choose to cruise the Nanticoke due to a lack of picturesque scenery, interesting ports of call, and marine facilities. A few miles inside the mouth of the river there is a considerable amount of fastland on the eastern shore of the estuary, and years ago the fishing villages of Nanticoke, Bivalve (a most appropriate name), and Tyaskin were founded. There are harbors and boat launching ramps at Nanticoke and Bivalve, but only the barest of marine necessities are available. The harbors and boat slips are used almost entirely by the commercial watermen of the area and the local residents who berth their pleasure craft there. Yachting and boating guidebooks and magazines virtually ignore the existence of the Nanticoke. It is, after all, primarily a commercial body of water. The numerous watermen who work out of Nanticoke, Bivalve, and Tyaskin reap a bountiful harvest of oysters, crabs, finfish, and terrapins at various times of the year. Presently there is a large seafood packing company situated inside Nanticoke Harbor, so that the watermen do not have to travel far to dispose of their catch.

As late as the early 1900s steamboats plied the waters of the river making regular stops at Tyaskin Wharf, located just inside the mouth of Wetipquin Creek, Vienna, Sharptown, and Seaford. The old wharf at Tyaskin is still in existence, but the channel of Wetipquin Creek is so badly shoaled one is apt to run aground even at high tide if care is not taken. Likewise, the main navigational channel of the river above Long Point is quite narrow, and a skillful captain is needed to maneuver a barge with a tugboat successfully from Tangier Sound to Seaford without running into a shoaled area. The channel is well marked with buoys, but I have nothing but admiration for the masters of the tugs that push the barges up the river at night. The tug-barge is covered with a multitude of different colored lights, and it gives you an eerie feeling to see this floating Christmas tree passing slowly and silently in the dead of night.

Marshland

Every few years you may hear that the Army Corps of Engineers plans to dredge the channel from Nanticoke to Seaford to eliminate the dangerous shoals, but the plan never seems to materialize. One of the major problems associated with a dredging operation of this magnitude is the disposal of the thousands of cubic yards of dredged spoil material that would be sucked up by the dredges. Disposal sites consisting of many acres each would have to be established along the river, and sufficient suitable land is just not available. I would venture to guess the captains who sail the river will have to watch the buoys carefully and play the tides for many years to come.

The area is rich in folklore concerning the many species you may come in contact with at various times of the year (especially those capable of adversely affecting you), and I have included some of these stories. Most folk stories concerning animals contain at least a small grain of truth. I have endeavored to give a scientific explanation for many of these tales. Some yarns simply do not have a basis in fact and have to be accepted or rejected solely on the basis of your personal opinion on the matter.

Throughout the book I have taken the liberty of relating some of the experiences I have had while participating in the panorama of the outdoors, especially along the Nanticoke River. Each time you go afield to observe nature, hunt, or fish, you experience something new, and this is really what makes all outdoors activities meaningful and worthwhile.

The study of natural history is a never ending process because our environment is constantly changing. I have witnessed such changes—albeit, nothing earthshaking—in the short time I have lived on the river, but over a long period of time these small, gradual alterations of the ecology of an area add up and create new environments. The organisms caught up in this never ending drama must adapt to the new conditions or perish. I am sure future observers of the natural history of the Nanticoke will find many things have changed since these words were written. It is hoped some of the observations reported here will be useful for comparison at later dates.

1

The Alpha and Omega of It All

SUSPENSE WAS BUILDING UP within me as I hauled in the fine-meshed net I had towed behind the outboard for approximately a quarter of a mile. Would "they" be there or not? I was concerned with the relative numbers of organisms that make up the all important plankton of the Nanticoke. Plankton means freely-floating organisms, and the abundance of these microscopic creatures found in a body of water is an overall index of the water's productivity.

I removed the plastic collecting-vial from the apex of the cone-shaped plankton net and held it skyward to make a superficial examination of its contents. I was encouraged as it seemed the sample was a bit more turbid than the ones I had taken during the past few weeks, and this cloudiness might indicate an increase in numbers. To be sure, however, the sample would have to be examined under a microscope.

My concern increased as I headed for my laboratory in the basement of the riverhouse and prepared a slide of the water and its contents. "Voila!", I uttered aloud as myriads of living things suddenly popped into focus. Copepods, diatoms, green flagellates, water fleas, and dinoflagellates were present in abundance. Even without making exact qualitative and quantative determinations, I was sure the overall numbers were at least ten times greater than in any of my previous collections. This dramatic increase in plankton species meant the food chain of the river had been activated—or, to put it another way, the principal ingredients of the "bread of life" were now available for all things great and small to feed upon.

The food chain of the river may be likened to a multilayered pyramid with each tier representing a different group of living organisms and different levels of energy. Close to the base of the pyramid is a broad band that represents the plankton. Untold numbers of organisms make up this segment, and a tremendous amount of energy contained within them is involved. As one surveys the different levels of the pyramid the higher tiers are found to be composed of fewer and fewer organisms and less and less total energy. There is also a constant progression to larger and larger creatures as the apex of the figure is approached.

The plankton of an aquatic environment is divided into two divisions—phytoplankton and zooplankton: plantlike and animal-like organisms respectively. Every food chain must start with the organisms known as the "producers." These creatures produce food from raw materials for the "consumers," and it is the microscopic phytoplankton that serves the vital purpose of producing the food (with energy incorporated in its molecular bonds) for all the other species found at higher levels on the pyramid.

All energy comes from the sun, and most members of the plantlike plankton use the green pigment chlorophyll to trap this energy and use it to convert the raw materials of carbon dioxide and water into carbohydrate foodstuffs. This, of course, is the basic biological process of photosynthesis which is necessary for life (at least as we know it) on earth. In addition to manufacturing food, the phytoplankton releases free oxygen into the environment as a by-product of photosynthesis. This oxygen is used by almost all living things to break down the molecules of food they have ingested and at the same time release the energy stored in the atomic configurations. Organisms capable of manufacturing their own food with the aid of solar energy are called autotrophs, while those that cannot do this (including all animals) are termed heterotrophs. The autotrophs of the world are the producers and the heterotrophs are the consumers.

The Alpha and Omega of It All

The phytoplankton is produced in trememdous quantities and is devoured by the zooplankton which includes such microscopic creatures as the water flea, cyclops, *Daphnia,* and other small crustaceans generally referred to as copepods. Being crustaceans, the copepods are closely related to such familiar animals as the crab, crayfish, and lobster. Copepods are gobbled up in large numbers by slightly larger animals such as fish fry, oyster spat, and worms which are in turn devoured by still larger animals. Eventually a perch, rockfish, catfish, or crab is produced which may end up providing the evening meal for a man, eagle, great blue heron, or raccoon. One can readily understand why it takes fantastic numbers of small organisms at the lower levels of the pyramid to produce a relatively few large individuals at the top.

This question of reduction in numbers as the apex of the pyramid is reached is related to the Second Law of Thermodynamics. This basic law of physical science states that when energy is transformed from one state to another there is always a net loss in usable energy units. The energy we acquire by eating a rockfish fillet is the same energy the phytoplankton trapped from the rays of the sun in the beginning. The First Law of Thermodynamics tells us that energy can neither be created nor destroyed. So, as the original solar energy was passed from consumer to consumer, along the line much of it was dissipated as heat or in some other unusable form. When the question is asked, "How many copepods does it take to make a rockfish?", the answer is, "a lot!"

A food chain can have no missing links. Any scarcity of individuals along the line will disrupt the flow of energy, and the numbers of animals that could be supported in the ecosystem are reduced, especially near the top of the figure. Population explosions, where too many individuals in one tier of the pyramid are produced, can likewise upset the delicate balance of nature.

In certain areas of the Gulf of Mexico the so-called red tide is not an uncommon occurrence. The water takes on a reddish color due to the presence of untold numbers of tiny dinoflagellates which are an important part of the phytoplankton.

These organisms produce a deadly toxin that can accumulate in the environment to such an extent massive fish kills occur. Man and other animals may be poisoned if they dine on fish impregnated with the poison.

Fortunately for us, we do not have the toxin-producing type dinoflagellates in the waters of the Chesapeake Bay system. When their reproductive processes run wild resulting in abnormally high numbers, however, certain other species of this flagellate in our waters produce what has been termed the mahogany tide. Many people along the Nanticoke have heard of the red tide and believe the mahogany type is the same thing. They become very apprehensive when, after a mahogany tide is noted in the area, large numbers of fish and crabs are noted laying dead on the beaches. These fin- and shellfish were not killed from a distinct toxic entity but died from suffocation due to an insufficient supply of dissolved oxygen in the water. The dinoflagellates had simply exhausted the oxygen supply in the river in performing their own respiratory functions.

The red or mahogany color of the water comes from a tiny red pigment spot each of the dinoflagellates has to help orient itself toward the sunlight when it performs photosynthesis. Sometimes these microorganisms are so numerous oysters and clams are affected. The flesh of mollusks takes on a red or pinkish tinge, and most people along the river will not touch it with a ten-foot pole. The flesh is, however, perfectly safe to eat as are the fish and crabs that were killed because of oxygen depletion.

As a matter of fact the reddish meat of the shellfish is caused by a concentration of the tiny red pigment spots the mollusk has filtered through its body. Regardless of all of this, shellfish having this abnormal coloration are almost worthless for the commercial market. It should be pointed out, however, that one should never eat any animal (reddish or not) taken from an area where the true red tide is present. Severe illness or death can result as the toxin is apparently one of the most potent known to science.

The Alpha and Omega of It All

The most important members of the zooplankton are the copepods. As noted, copepods are crustaceans and considered by some authorities to be the most abundant group of animals on the face of the earth. They occur in both fresh and salt water and become so numerous at times a count of three million may be made in a cubic foot of water. Thirty different species of copepods have been found in the waters of the Chesapeake, but ninety-five percent of the fauna are composed of only five different types.

Copepods are eaten for their energy by virtually all larger animals in the Bay and its estuaries. However, oysters and clams do not feed directly on these tiny crustaceans. Instead, they feed on the decomposed organic matter on the bottom which results mainly from the dead bodies of copepods that did not find their way directly into the digestive systems of larger animals. Most of the food of shellfish comes from the phytoplankton which is filtered through their bodies in an unending current. When copepods are in abundance in an aquatic environment, it is fairly safe to assume the overall food chain is functioning properly.

I once used a quantity of copepods collected in my plankton net to conduct an unscientific experiment to determine the number of prey organisms a predator would consume in a given time. I placed an estimated five thousand copepods in a tank with a silverside minnow about one inch in length. At the end of a twenty-four hour period not one of the crustaceans could be detected on the numerous microscopic slides I examined. No visible difference in the size of the minnow could be detected after its daylong orgy. Apparently the copepods were ingested, digested, assimilated, and egested rather rapidly.

Because they are eaten in such great numbers, copepods must be able to reproduce rapidly. When conditions are optimal (i.e., there is an abundance of phytoplankton), the number of offspring produced is colossal. One copepod expert has estimated that if all the offspring resulting from a single pair lived and reproduced, and so on for a period of one year, the face of the earth would be covered with copepods to a depth of six feet! We can offer thanks that only a small fraction of these crustaceans is not eaten as soon as it hatches from the eggs.

During the warm months of the year ova are produced continuously and hatch almost immediately after they are laid. As the waters grow colder in the fall, egg production is slowed down and some of the eggs are deposited with thick shells around them. These thickened coverings prevent the eggs from hatching until the water of the river warms in the spring. Many copepods overwinter in the egg stage, but some pass the harsh cold months in the adult form.

Like other members of the plankton, these crustaceans are largely dependent on the winds and tides for their locomotion, but within their own little bailiwick they are capable of limited self-induced movement. Five pairs of swimming legs are present on the fused head and thorax. In addition, one of their two pairs of antennae or feelers are frequently used as sculling oars. The antennae undoubtedly have certain sensory functions like odor detection, but it is thought their primary function is to jerk the tiny beast along in conjunction with the other appendages. The antennae in some copepods are extremely long, and this condition has caused many biologists to refer to them as "longhorns."

Other appendages about the head region serve to swish small microorganisms such as dinoflagellates and diatoms into the mouth and gullet in a never ending flow. In many there is a filtering system incorporated into the mouth which aids in the selection of the goodies. Copepods also utilize organic debris as food, and the course of the ingested nutrients through their alimentary tracts is easily followed by viewing through the microscope. The animals are more or less transparent, and globules of fat and the red pigment spots of the phytoplankton are easily observed.

It is interesting to realize population explosions of certain copepods coincide with the hatching of the eggs of many species of fish. The young fry are thus assured of an adequate food supply during the critical days of early development. The spawning runs of the rockfish up the Nanticoke each spring

Green algae longhorn

are timed to the first massive hatch of copepods of the season. One way to determine if a "good rock year" is in the offing is to check the copepod level of the river about the middle of April.

The coordination between copepod numbers and fish-egg hatching is not mere coincidence but the result of millions of years of evolution of an entire ecosystem. Through countless ages the links of the food chain were tested, rejected, accepted, or modified until a system evolved that was best suited to the members of the community under a given set of environmental conditions. It seems obvious to me any disruption or malfunction of any tier of the pyramid can bring the whole structure crashing down into biological ruin.

The next time we (the Omega) sit down to enjoy a broiled rockfish we should give thanks to the plankton (the Alpha).

≈ 2 ≈

Sea Nettles and Jug-Stoppers

AS THE DIVERSIFICATION of living things on earth shifted into high gear, perhaps two billion years ago, the first truly multicellular animals to appear in the scenario were the jellyfish and their relatives. Many of these primitive beings remain relatively unchanged today. As the flatworms diverged from the jellyfish, there evolved three primary cell layers from which the tissues and organs of the body would develop; and this plan of development was to be used by all the other multicellular creatures that would follow in the eons ahead. The ancestral jellyfish were destined to exist forever without the important third bank of developmental cells which turn into such structures as bone, blood vessels, kidneys, and muscles. In place of these cells in the jellies is an acellular, gelatinous-like material that is ninety-nine percent water. Though the jellyfish have remained very primitive in most of their anatomy, certain specialized cells and structures did develop. Jellyfish also may have extremely complex life cycles which have evolved in response to environmental selective pressures molding them into a group of the animal kingdom which is well suited for its particular place in the environment.

One feature of the true jellyfish which is singular to them is the highly specialized stinging cell used in the capture of their prey. The stinging cells (properly called nematocysts) of most jellyfish are of no real threat to man's life or limb, but in some species it's a different story.

Sea Nettles

The sea nettle that frequents the Nanticoke and the rest of the Bay during the warmer months of the year and reaches its greatest abundance during late summer and early fall is a dangerous species. When the skin of a human being makes contact with one of these translucent blobs of protoplasm an intense pain and fiery itching sensation always result. I asked an old man loafing about the pier at Bivalve one day what was the best or most effective thing to do when one is stung by one of the long, trailing tentacles of the sea nettle. He replied candidly, "About the only thing I know to do is to first grit your teeth as hard as you can and then holler as loud as you can. Hit won't do no good, but you gotta do something."

It is almost unbelievable that this primitive animal can pack such a wallop when it only brushes against bare skin, but there are untold numbers of witnesses in many parts of the world who can testify to the severity of the reaction caused by this contact with a sea nettle. The nettle enjoys a wide distribution in the marine waters of the world, and you can make this unfortunate contact from the mid-Atlantic states to New England, across the Atlantic Ocean to the Azores and on to West Africa, around the Cape to the Indian Ocean, and on into the Pacific Ocean to Japan and the Philippine Islands. The nettle reaches its greatest abundance along the shores and estuaries of the Delmarva Peninsula, and, for some reason, population explosions will occasionally occur in the Chesapeake and Delaware bays. I have had residents of these areas tell me that during these times of abundance, the jellies may be so thick one can walk across the surface of the water and not get his feet wet. When jellyfish occur in such stupendous numbers, they clog the waters and become a menace to the crabber, fisherman, waterman, and swimmer. Crab pots and fish nets may become so entangled with the numerous bodies of these gelatinous demons that the removal of the catch becomes tedious or downright dangerous to say the least. Many part-time net fishermen along the Nanticoke will gather in their gill nets around the first of August and not reset them until the jellies

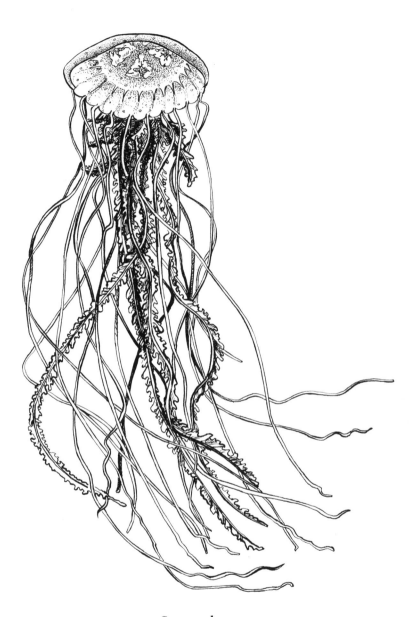

Sea nettle

have abated, late in the fall. This midsummer removal of the nets is also tied in with the abundance of crabs in the river at that time. Crabs have the nasty habit of making their way into the nylon meshes of the net to nibble at the trapped fish, destroying much of the catch. Many times, while rustling the fish, the crabs also become entangled in the net, and it does not take an angry crab long to tear a section of expensive net to shreds as it attempts to cut its way to freedom.

During the times of sea nettle proliferation, it is rare indeed to find an individual who is bold enough to venture into the waters to enjoy a refreshing dip. In some years swimming is almost completely curtailed, with only those who have access to a beach protected by a jellyfish-proof screen or small-meshed net able to enter the water. These screens, however, seem to offer little protection after they have been in the water for a few weeks. Wicomico County owns and maintains an excellent picnic-beach recreation area in conjunction with the harbor at Bivalve, (Cedar Hill Park), and each year in June workmen place a jellyfish curtain in the shallow water off the white sandy beach to protect the bathers. By the middle of July each year the protective screen has developed many holes large enough for the slippery jellies to wriggle through, and, in reality, the once protected area is effectively turned into a jellyfish corral. Unsuspecting users of the beach have frequently entered the water surrounded by the enclosure believing it was free of the dreaded nettle only to find themselves in a veritable hotbed of stinging tentacles. A marine policeman stationed at the harbor once told me a child was stung so severely several years ago that he lapsed into a state of unconsciousness and drowned in two feet of water.

The stinging cells of these animals are amazingly complex, and one cannot help but marvel at a natural system which will permit a primitive group of animals to possess such a sophisticated and refined weapon. Hundreds—even thousands—of these threadlike structures are located on the long tentacles and around the single opening to the body which serves as both a mouth and an anus. Some of these cells are unarmed and serve to entangle the prey in a jungle of filaments, but

Sea Nettles and Jug-Stoppers

another and more numerous type of nematocyst is sharp-pointed and readily enters the body covering of almost any animal coming in contact with it (the diamondback terrapin and other river turtles being exceptions). Each of these armed cysts consists of a tightly coiled, hollow filament contained in a capsule filled with venom. It is neatly booby-trapped by a restraining lid that is tight fitting and equipped with a sensory hair-trigger mechanism protruding from it. When the tentacle brushes against the skin, the trigger activates the lid of the capsule, and it flies open. In an instant the coiled poison dart is released with considerable force. The dart is hollow, and there is enough power stored in the coil to drive the sharp pointed apex into the skin. The venom is injected hypodermically, and, as the old man said, about the only thing you can do is grit and holler with all your might. This procedure seems to work about as well as any other course of action.

The symptoms resulting from the virulent toxin injected into the body vary greatly with the individual. If you are lucky you will get away from an encounter with this monster with only a period of intense pain and itching. If, however, you are not so lucky and happen to be especially sensitive to the poison, general symptoms may develop in the form of vomiting, frothing of the mouth, muscular cramps, respiratory difficulties, delirium, convulsions, or even death. If a tentacle happens to come in contact with the eye, the resulting pain has been described as a sensation something like what you would expect if a red-hot nail was driven into your forehead. Jesse Bloodsworth of Crisfield relates an encounter with a nettle several years ago. Bloodsworth was working the patent tonging device for oysters on a workboat, and as the tongs were swung on board a sea nettle lodged among the oysters slid free and struck him full in the face. The pain in his eyes was excruciating, and he was blinded instantly. He reeled backwards in agony and went over the side into rough seas. He surfaced a few yards off the boat and with blurred vision just managed to grab a line which had been thrown to him by his companions. As he was being hauled back aboard he suddenly realized the pain had lessened considerably, and his vision seemed to be improving. Apparently the water had served to wash away, or at least dislodge, many of the offending nematocysts embedded in his eyes. He finished the day's work in spite of severe headaches and irritated eyes. His normal vision did not completely return until three days had passed.

For decades scientists have attempted to pinpoint the chemical nature of the jellyfish venom, but, so far, have been largely unsuccessful. However, it is known that the material consists of several different toxins which are undoubtedly proteinaceous in nature. This protein base of the substances accounts for the fact many people become hypersensitive to the stings as a result of repeated exposures. Such a sensitized individual may well lapse into anaphylactic shock and lose consciousness when subsequently stung in a manner similar to what may occur as a result of repeated bee and wasp stings. One can speculate that this is what happened to the child the policeman reported as having drowned in the shallow water of Cedar Hill Park.

Fortunately, most people suffer only the localized symptoms of itching, burning, and pain. In addition to gritting and hollering, numerous methods of treatment of localized sea nettle sting have been suggested. One of the most common remedies one hears about along the river is to rub the affected area with sand, and I can personally attest to the partial efficacy of this procedure. Some Nanticoke residents insist the use of dark sand is more effective than the white variety. Other topical materials that have been advocated for therapy include household ammonia (it seems ammonia is suggested for all types of stings and bites so there must be something to this time-honored treatment), baking soda, alcohol, beer, sugar, olive oil, kerosene, gasoline, and urine. The nitrogenous urinary product urea will quickly convert to ammonia when exposed to the air, and this undoubtedly accounts for urine being a frequently suggested treatment for bites and stings.

The Maryland Institute of Marine Science suggests the application of commercial meat tenderizer for nettle sting. It is believed the tenderizer acts as a poultice and draws out the

Sea Nettles and Jug-Stoppers

nematocysts and the poisons. As the tenderizer consists mainly of enzymes which digest proteins, it is more likely effective because the nematocysts are digested or dissolved away, and the protein-based venoms are denatured. Unfortunately, there never seems to be a jar of meat tenderizer handy whenever I am stung.

It should be noted the stinging cells are just as effective out of water as in it. A first-class hot foot will result from treading barefooted on a beached nettle, and a specimen should never be picked up for examination with the bare hand.

The sea nettle is only one of four common species of jellyfish one is apt to encounter in the marine and brackish waters of the eastern United States. Although there are many other species of this type of animal in our waters, they are largely unnoticed by the untrained eye or casual beachcomber. Of the four, we can take comfort in the fact the sea nettle is the only one capable of causing discomfort in human beings. The venoms of the others are highly effective against their normal prey animals, but, for some unexplained reason, cause little or no adverse reactions in humans. The sea nettle's venom is obviously different in its chemical nature and composition.

The nettle is readily distinguished from the others by the twenty-four, long tentacles that trail along behind the pulsating, four-inch umbrella or bell. The common moon jelly may have an umbrella or body up to eighteen inches in diameter with tiny tentacles forming a short fringe border around the lower edge. The so-called lion's mane may measure up to fourteen inches across the bell and has several large, armlike tentacles attached to its body opening. The red winter variety of jelly is usually about eight inches in the beam and has an arrangement of tentacles in eight large clusters on the underside. It is a first cousin to the sea nettle, but the different setup of the tentacular processes is distinct and makes it easily distinguished. It should be mentioned that the identifying characteristics of the four jellies just described are easily visible, and it is not necessary to come eyeball-to-eyeball with one of the four to determine its identity. Anyway, I can think of no reason why anyone would want to fondle a jellyfish of any type, but curiosity has driven people to do strange and unusual things.

The sting of the sea nettle is frequently compared to the potency of the sting of the Portuguese man-of-war. The man-of-war, or bluebottle as it is frequently called, is not found in the latitude of the Chesapeake but in warmer southern waters of the Atlantic and the Gulf of Mexico. Actually the man-of-war is not a single jellyfish but a colony of several different types of jellylike animals attached to a wedge-shaped float filled with gas. The float is bluish in color and serves as a sail by which the wind blows the colony from place to place. To some romantics the float appears to resemble the sail of an old-time sailing vessel, and, as the "ship" houses cannons loaded with potent nematocysts, it has been dubbed the man-of-war. Some members of this pelagic colony are responsible for the perpetuation of the species and other members (the ones loaded with nematocysts) function to ensnare and capture the prey. Some biologists claim the bluebottle packs the most powerful sting of any marine animal found in the waters adjacent to this country. Other experts refute this and claim the sea nettle is just as bad (if not worse). I have been stung by both types and am inclined to go along with the group of jellyfish pros who give the nod to the man-of-war. Once while swimming in the Gulf at the mouth of the Rio Grande at the tip of Padre Island in southern Texas, a wave washed a bluebottle into my midsection. The pain was overwhelming, but after a short time it seemed to pass away, and I had the euphoric impression I was floating away, upwards toward the sky. Within a short time I returned to earth (or water as it was) and the pain reintensified itself. Fortunately, I was not far from shore and the water was not deep, so I was able to struggle back to the beach, albeit not in the best of condition. One of my companions sprayed the reddened area on my abdomen with a sunburn concoction containing the local anesthetic benzocaine, and somehow I managed to survive. During the period of the treatment two surf fishermen across the river in Mexico kept shouting at the top of their lungs, "Peese on it. Peese on it." Apparently the

Sea Nettles and Jug-Stoppers

knowledge of the effectiveness of urine in alleviating the pain of jellyfish stings enjoys international reputation, and is not confined to first aid treatment along the Nanticoke.

When it comes to the question of which jellyfish in the world is the most dangerous, there seems to be little doubt that the dubious honor goes to a creature known as the sea wasp. This jellyfish is an inhabitant of the waters of northern Australia, the Philippines, and the Indian Ocean. Though barely four inches in size, numerous deaths from its sting are recorded each year. One medical authority states a person may die from its sting in as little as three to eight minutes. During a recent twenty-five year period, in the waters off Queensland, Australia, sharks claimed the lives of only thirteen victims while sixty persons were killed by sea wasps. Recreational and commercial usage of our waters would be drastically curtailed if, by some ecological quirk, the sea wasp were ever introduced into this part of the world and were able to establish itself.

Frequently after a bad jellyfish season you will hear a demand from irate citizens that a jellyfish-control program be instigated by some governmental agency. They reason that if we have such programs for the abatement of mosquitoes and greenhead flies, why not one for the sea nettle. With his almost unlimited scientific capability, I suppose man could, after spending millions of dollars for research, devise some sort of semieffective control or eradication program for the sea nettle. But are these creatures of any benefit to mankind?

Generally speaking, jellyfish are of little or no direct economic importance in this country, but in parts of the Orient they are relished as food. In that part of the world jellyfish of various types may grow considerably larger than they do here (one species measures up to seven feet in width). These jellies are cut into strips, treated with salt and alum, hung up to dry on long lines, and then stored. When dehydrated in this manner, the tissue (what precious little is left after the ninety percent water is removed) must be reconstituted before use. After the water is replaced by soaking, the jelly is fried in deep fat and is said to resemble pork cracklings in taste. If the jellyfish are boiled without first undergoing the drying out process, they taste like good old-fashioned tripe, according to one globe-trotting gourmet. I have often thought that the next time I ran onto a mess of jellyfish either in my net or washed up on the beach, I would prepare them (I have always had a fondness for cracklings) to satisfy my taste curiosity. So far, however, the *right* occasion has not presented itself, and I confess I have not yet experienced the delight of a dish of jellyfish.

One other use of the sea nettle has been brought to my attention by my neighbor on the river, Lou Griffin. Griffin relates that when he was a boy it was a common practice for the youngsters of the area to dry jellies in the sun for several days and pulverize what was left into a powder. This fine material was placed in a pill box and taken to school where it was periodically utilized as a prankster's sneezing powder. The dried nematocysts apparently did a bang-up job of irritating the respiratory tracts of the students and teachers alike and precipitated many outbursts of uncontrollable sneezing that effectively disrupted the classroom routine.

Even though everyone who has had anything to do with jellyfish understands their nuisance value, we know surprisingly little of their place in the overall ecology of the seas, bays, and estuaries they inhabit. Most certainly they are a part of the all-important environmental food chain, but just how important they are to the overall strength of that chain is unknown. It is a fairly safe bet that these primitive organisms are more intimately involved in the interaction of life in the marine environment than most people suspect. The balance of life in any ecosystem is a very delicate arrangement of species, in harmony with the existing environment, that has emerged only after millions of years of evolutionary trial and error. To tamper with any of the threads in this fragile web of life is to court ecological disaster. Until more knowledge is gained of the role the jellyfish play in this drama, no artificial control program should be instigated. Human intruders in their bailiwick are just going to have to "grit and holler" as they pay the price for the invasion of another species' ecological domain.

Sea Nettles and Jug-Stoppers

Jug-Stoppers

Another type of animal frequently encountered along the Nanticoke is a member of a group of animals closely related to the jellyfish and their allies. These smallish creatures are variously referred to as comb jellies, sea walnuts, comb bearers, and jug-stoppers. These animals are more advanced, from an evolutionary standpoint, as the layer of jelly found in their distant cousins is replaced with a true cellular layer typical of the other multicellular members of the animal kingdom.

Comb jellies along the Nanticoke are sometimes called jug-stoppers because of their approximate four-inch oval shape which resembles (at least to some people) a corking device of some sort. Fortunately for us, the jug-stopper differs from jellyfish in another very important way—it lacks the terrible nematocysts that are characteristic of all true jellyfish. The stinging cells are replaced by adhesive cells called colloblasts which serve to capture food. The colloblasts produce a secretion that aids in entangling food organisms in a contractile, spiral filament. Recent studies have shown that, indeed, these adhesive cells contain a venom that intoxicates its victims, but it is not injected into the body. As far as I can determine, there has never been a report of a human having been injured by contact with one of these animals.

Jug-stoppers and other members of this group are almost transparent though some may exhibit a pinkish or greenish tinge. They are called comb jellies because they all have eight rows of comblike structures on the sides of their bodies that beat in unison and propel the animal through the water in a manner similar to an ancient oar-powered galley. A specialized organ of equilibrium or gyroscope is located on the top of the body and serves to let the beast know in a general way which side is up. This organ is correctly called the statocyst and consists of a ball of sandlike material enclosed in a cyst or vesicle. Projecting into the cyst are numerous fine hairs attached to primitive nerve fibers running throughout the body. If the animal leans to port, the ball rolls in that direction and touches the sensitive hairs. A signal to come about to starboard is then relayed throughout the body. The whole thing reminds me of the "Tilt" mechanism on an old pinball machine I used to own.

Many comb jellies have the ability to reflect brilliant changing colors as they move through the water with the combs beating. If they are near the surface, light strikes the filaments of the comb and is refracted into the colors of the rainbow. In addition, many comb jellies have the eerie capability of generating phosphorescent light at night. Supposedly this bioluminescence is useful in attracting food organisms within range of the colloblasts.

I had known since my first course in invertebrate zoology in college that these animals were capable of generating light, but this fact had been lost in the recesses of my mind until one June night as I waded out to check the catch in my fish net in front of the cottage. The tide was ebbing strongly, and I was surprised to notice that the water seemed to be glowing as it swirled around my waist. It was a strange sensation to say the least—to be standing in water that seemed to be electrified. Closer observation revealed the light was not distributed evenly throughout the water but emanating from distinct entities under the surface. I procured a large-mouthed jar and captured several of these blobs of light as they drifted by, and then proceeded to my small laboratory located in the basement of the cottage. The luminescence continued after the objects were confined to the jar, and as I waded ashore with my biological lantern held on high, I somehow felt like Diogenes, two thousand years ago, wandering around looking for an honest man. The glow disappeared in the light of the laboratory, but when the contents of the jar were emptied into a pan, it became immediately obvious to me what these strange organisms were. I took the pan with the creatures in it out into the dark backyard, and instantly they commenced to emit light as strong in intensity as before. They continued to glow throughout the night. But, in the morning they were as dead as a run-down battery into which no energy had been returned as it discharged its power. That night I gazed out over the river

Sea Nettles and Jug-Stoppers

and could see long, slender bands of light that stretched as far as the eye could see.

Watermen I have talked to are not aware that one of the favorite dietary items of the jug-stopper is oyster spat or larvae. Jug-stoppers have population explosions like the sea nettle and may be found in the river at a density of up to fifteen individuals per cubic meter of water. Studies have shown that each jug-stopper is capable of catching and digesting twenty-two thousand spat in a twenty-four hour period. Therefore, if periods of overpopulation with these predators happen to coincide with the hatching of the spat, obviously considerable damage can be done to that year's oyster hatch.

But, when the jug-stoppers become that plentiful the situation does not exist for long. Nature has a way of taking care of abnormal numbers of any given species. Jug-stoppers are voraciously devoured by many species of fish. A small white perch is known to be able to swallow a dozen jug-stoppers at a single meal. There are lots of white perch in the river, so the abnormal population of jug-stoppers is quickly and effectively reduced to its usual size.

These primitive animals that exhibit complex specializations fascinate me. The generation of bioluminescence is infinitely more complex a procedure than the generation of electricity in an oil or nuclear powered generator and the subsequent transferring of it to an incandescent light. To me this phenomenon ranks as one of the great marvels and mysteries of the natural world. I don't suppose I will ever figure out why such a sophisticated system evolved in so primitive a creature and not in an animal such as myself. In any event, pondering this sort of thing gives me something to do while sitting on the banks of the river on warm summer nights watching the glowing water in the distance.

~ 3 ~

The Horseshoe Crab, A Living Fossil

AS SPRING slowly changes into summer and the waters of the Nanticoke gradually become warmer and saltier, the horseshoe crab (sometimes called king crab) makes its annual invasion from the ocean and the mouth of the Bay. (Although the name king crab usually brings to mind the succulent sections of the legs of the Alaskan king crab you can purchase in the supermarket, you should take care not to confuse this species with the living fossil.) Hormones begin to flow in the bluish fluid of its body, and instincts that have governed the behavior of this animal for over four hundred million years take charge and tell it to move shoreward and begin the process of perpetuating the species.

Frequently a large female will be seen moving along the shore with several smaller males in attendance. Sometimes one of her consorts will hang onto the female's abdomen by means of a pair of clasping like organs located at the ends of the first pair of large legs. This male is probably the one that will sire the next generation, as he has proven stronger and more determined than the lesserlings.

With the successful male riding atop, the female clumsily moves out of the surf and well onto the beach. She then proceeds to dig out a shallow nest into which two to three hundred eggs are deposited. The male is then pulled over the nest by the female, and sperm cells are ejaculated over the eggs. Having fulfilled their biological duty, the pair leave the

The Horseshoe Crab, A Living Fossil

nest and head back for the water of the river or ocean. Waves quickly wash sand into the nest sealing it into an incubator.

Eggs are always laid at what is called "spring tide," which is usually high and coincides with the appearance of the full moon. Then, the fertilized eggs have about two weeks to complete the initial stages of their embryology. By the time the next spring tide rolls in, they are ready to be washed out to sea. The young crabs are completely at the mercy of the elements, and undoubtedly many are stranded on the sand and perish before they reach the water.

Older crabs are said to be possessed with positive geotropism, which means they are attracted to the pull of gravity. If they are stranded on the beach as the result of a storm, they always head downhill and most often this leads them to the water's edge. Once I found an old sow crab, as they are sometimes referred to along the river, stranded on the beach at Tyaskin. She was slowly moving toward the water down the very slight incline of the beach. To test the theory of positive geotropism I covered the crab's four eyes with a small piece of opaque masking tape and secured a wide piece of driftwood laying on the beach. I constructed a ramp leading away from the water's edge at an angle that was a little greater than the slope of the beach. When I placed the large sow on the board, she headed downhill in response to the pull of gravity. When she crawled off the end of the ramp she continued onward away from the force of gravity for about a foot. She then stopped and appeared puzzled, but soon she slowly made a U-turn and headed downhill again. As she reached the water's edge I removed the tape from her eyes, and she disappeared into the depths.

The horseshoe crab, or king crab as it is frequently called, is not really a true crab at all but a member of that great group of animals without backbones known as the *Arthropoda* or jointed-legged animals. The *Arthropoda* includes, in addition to the horseshoe crab and the true crabs, such familiar species as ticks, mites, scorpions, spiders and insects. But, no matter where zoologists classify the horseshoe crab, it most assuredly is biologically adapted for survival.

When we look back to the fossil record of the geological period in the history of the earth known as the Paleozoic era, it is striking to note that of the thousands of primitive animals that evolved and existed some four to five hundred million years only one has endured to the present time in an unchanged condition—*Limulus,* the horseshoe crab.

During this prolific expanse of time, thousands of different animals evolved. But, when the environmental conditions changed as the era gradually came to an end, the existing species were subjected to the awesome forces of natural selection; and to the survival of those best suited to cope with the new environment. It was a matter of change or perish.

Many types became extinct, and others of a more plastic genetic makeup were able to change and evolve into many of the forms with which we are familiar today. Only *Limulus* was able to ignore the forces of nature and time, and continue on, unmodified and thriving, to the present day. No one really knows why *Limulus* was able to prevail, but one explanation (as good as any of the others) is the horseshoe shape of his shell.

This horseshoe-shaped shell which covers the animal's combined head and thoracic regions certainly offers excellent protection against possible predators, and the smaller abdomen is also encased and protected by a heart-shaped shell bearing numerous spines and projections along its edges. Trailing backwards from the abdomen is a long, fearful looking, pointed spine called the telson which aids in navigation both in the water and on the beach. As previously noted, on the upper surface are two pairs of unblinking eyes, and on the underside are numerous appendages to assist the animal in the capture of food, crawling about, and burrowing around on the bottom. The breathing organs, constructed like the pages of a book, are also located on the under surface.

Horseshoe crabs may reach an overall length of almost two feet and are dark brown in color. No one knows for sure how long a horseshoe crab will live; but when you consider they do not become sexually mature and capable of reproduction for about nine to ten years, you can feel sure they live at least nine

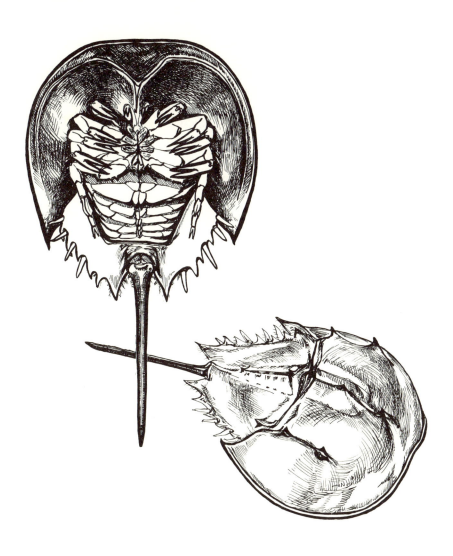

Horseshoe crab

or ten years longer after reaching puberty. Normally they move slowly along the bottom where they feed most actively at night on clam worms and many other small animals they may encounter. Contrary to popular belief, the long, menacing, telson is not an offensive or defensive weapon. Many people in this area, however, adamantly believe the telson is a poisonous spear similar to the dangerous tail of the stingray.

Once while surf fishing on Assateague Island I was startled by the sudden scream of a bronzed young woman who was frolicking in the surf with a towheaded toddler. Her sunburned face paled as she suddenly swept the child into her arms and scrambled madly from the sea to the safety of the beach. When I approached her and asked what had occurred to evoke such a reaction, she replied she had been brushed by a monster in the shallow water—a beast with a terrible spine sticking out of its rear end.

I waded into the surf and discovered a large, floundering, female *Limulus* struggling against the breakers to make her way to deeper water. I grabbed the crab by her tail and placed her bottom side up on the beach. The woman was absolutely amazed that I would be so foolhardy (or so brave) as to tangle with such a "dangerous" creature. She would come no closer to the crab than ten feet and turned a deaf ear when I tried to assure her the animal was absolutely harmless. As we watched, the crab rotated its telson in an attempt to drive the tip into the sand. The woman was convinced it was trying to implant its deadly weapon into her flesh or that of her offspring. Finally the apex of the telson lodged in the sand and enough leverage was gained so the other appendages also touched the ground. With a mighty heave the animal managed to right herself and maneuver the head toward the water (downhill). Slowly but surely she made her way back to the edge of the surf and was swallowed up by the foam. The characteristic track made in the sand by the dragging telson was quickly obliterated by a roller, but the impression left in the mind of the tourist from Philadelphia will probably last as long as she lives and become the basis of yarns she will undoubtedly spin to her grandchildren.

The Horseshoe Crab, A Living Fossil

The young woman in the episode just described called her "attacker" a devil crab, but I have never heard that name applied to the horseshoe crab since. Some people incorrectly call it a mole crab, but this appellation is more generally reserved for a type of small, true crab that uses its molelike front legs to burrow rapidly into the sand at the edge of the surf. Early residents of the Delmarva Peninsula used the shells of larger specimens as wash basins and containers of various sorts and called them pan crabs.

Whatever name is applied to *Limulus polyphemus*—its only true scientific name—this armored arachnid is found in the Atlantic and its tributaries from about forty-five degrees north latitude to the Gulf of Mexico. Its long evolutionary history is recorded as fossils in the rocks of various geological formations throughout the world. In fact, when the fossilized remains of the first bird, *Archaeopteryx,* were uncovered in Germany many years ago, the rock also contained a well preserved *Limulus.*

On the Nanticoke River I have found them as far north as Penknife Point, and I suspect they may even roam as far upstream as Vienna at certain times of the year. Dr. Tom Jones of the Salisbury State College Biology Department tells me they are very common at Horn Point on the Choptank River near Cambridge. They are frequently found on the Atlantic beaches of Maryland, New Jersey, and Delaware and move into Delaware Bay in tremendous numbers. Of course, they also reside in the Chesapeake as far north as the Bay bridges or farther.

Most marine biologists agree the numbers of horseshoe crabs have declined during this century. However, there are reports from years ago of the huge numbers that frequently washed ashore in southern New Jersey at the mouth of Delaware Bay. Farmers in the Cape May area would rake the numerous crabs into small mountains and allow them to decompose under the hot sun for a period of time. When the crabs were "ripe," they would be carted away and ground up to be used as feed for chickens and hogs.

Many poultrymen were convinced something in the odiferous concoction induced their chickens to lay more and larger eggs. Other chicken and some pig farmers found the aromatic mixture gave the pork or chicken meat a strong pungent taste when cooked and thus rendered it unfit for human consumption. There is no indication eggs laid by *Limulus*-enriched chickens acquired an abnormal taste or smell.

Apparently these crabs were so plentiful around the turn of the century, they were also ground up in enormous numbers and used as a nitrogen-rich fertilizer for garden crops. The shells of all arthropods are made up of a very unique material known as chitin which is essentially a nitrogenous compound. It seems logical that the shells of the crabs would offer an excellent source of this element essential for protein production and other important constituents of living protoplasm. There is division of opinion among the experts as to the role this harvesting of the crabs for commercial endeavors played in the decline of the overall population. I am inclined to believe that these operations had little to do with the reduction in numbers of the species. I suspect most of the individuals collected off the beaches for commercial use were crabs that had already accomplished the act of reproduction.

If biologists think the numbers of horseshoes are less than they were in bygone years, this fact does not seem very apparent. So many of the armored denizens littered the sand at Rehoboth Beach, Delaware in 1979 that the city fathers of this resort community approved the hiring of a full-time worker to pick up crabs during the 1980 tourist season. Apparently, the solons did not want to take a chance in offending the dollar-laden vacationers with the smell of rotting crabs or perhaps evoke crabphobia among those who believe the beasts to be extremely dangerous.

A possible new threat to the horseshoe has arisen in recent years with a revitalization of the ancient practice of eeling. Watermen of the Chesapeake have always ensnared eels in great numbers to be used primarily as bait in their blue crab pots or traps. Not too long ago it was discovered (or redis-

The Horseshoe Crab, A Living Fossil

covered) that extraordinary bait for eels was a female *Limulus*, especially so if she were laden with eggs. Something in the makeup of the female (and to a lesser extent in the males) and her eggs is irresistible to this snakelike fish. These fish are being sought in increasing numbers by the watermen of the Chesapeake to supply a seemingly insatiable market in the Far East and Europe. In these areas eels are considered to be gourmet food, and in certain regions of this country they have a limited market where certain ethnic dietary habits continue. Many watermen are finding it more profitable to forget the blue crabs during the first part of the summer and go after eels full time. In pursuing this endeavor they need to procure all the horseshoe crabs they can to stuff into their eel pots.

Oyster and clam dredges are now utilized to dislodge the horseshoes from the bottom of the Chesapeake and its tributaries, but supply never seems to catch up with demand. The eeling waterman is now turning to the surf clammers in Ocean City, Maryland and elsewhere along the coast to provide the additional bait that is required. It was reported in 1979 a "good-sized" female *Limulus* (packed with roe) was worth fifty cents to the clam dredger. The eeler can cut such a specimen into enough bait to serve five or six eel pots.

This newfound interest in the horseshoe has turned out to be a bonanza for the clammer. Horseshoes are a known enemy of clams, and it has been estimated that one horseshoe crab is capable of destroying approximately one square foot of clam bed each day as it forages and scratches around in the bivalve nursery. Clammers would formerly routinely smash to death any *Limulus* they brought up in their dredges or encountered otherwise. Now these despoilers of the clam beds are carefully placed in a holding box for sale to the anxious eeler.

Despite the large size a horseshoe crab may attain, there is precious little about the beast most humans would consider edible. However, there is a small knot of muscle tissue at the base of the telson which serves to operate this organ, and a narrow strip that runs up the back of the shell that have been eaten by gustatory detectives. Those that have eaten these parts of the crab report them to be delicious, resembling lobster in taste. Considering the large numbers of crabs that would have to be caught and cleaned to attain enough meat for a meal, it seems quite unlikely *Limulus* à la Maryland will find its way to the menus of Delmarva or elsewhere. This, coupled with the fact most individuals would reject horseshoe crab dishes on aesthetic grounds, makes it improbable that watermen will devote any part of their operation to the taking of horseshoes for the table. There is far too much more money in eels. However, as everyone knows, the fickleness of human nature is frequently revealed in the ever changing dietary preferences of people. A hustling seafood marketeer could think up a fanciful name for our armored friend—say, "Fortified Maryland Crab"—and a new food craze could start. As far as I can determine, no one has ever tried preparing a soft *Limulus* in the way soft shell blue crabs are prepared and prized for the table. The horseshoe crab periodically sheds its outer covering of chitin in much the same manner as the blue crab and is likewise vulnerable to predation while the new shell is hardening. For a few days after the shell is shed the body is soft and pliable and, in the case of the blue crab, the entire animal is consumed with gusto after being rolled in flour and deep fried. I am going to have to try fried, soft king crab the next time I run into one on the river.

Putting these contributions of *Limulus* to the betterment of mankind aside, it should be pointed out the most important impact of these animals on our daily lives may come from scientific research. The blood of *Limulus,* as mentioned earlier, is blue due to the presence of a substance known as hemocyanin. Researchers have found the hemal fluid to be of great value in their investigations into the basic functions of the nervous and circulatory systems of higher animals. Other scientists have shown certain cells that float around in the bluish plasma to be of tremendous value in specialized tests designed to detect the presence of certain endotoxins or poisons in humans.

Finally, *Limulus* has always been of great interest to biologists, especially those interested in the study of evolution. The

The Horseshoe Crab, A Living Fossil

fact it has been able to endure in an unchanged form for some four hundred million years is a staggering one in the mind of the evolutionist. In evolution, the primary assumption is that change is brought about in response to environmental pressures—and here is a creature whose existence boldly defies this fundamental law of nature.

There is no doubt the horseshoe crabs evolved from a large group of the jointed-legged animals known as trilobites which flourished in great numbers on the bottoms of primeval oceans some four to five hundred million years ago. Trilobites readily fossilized and are frequently found by collectors when they examine rocks dating from the geological era in which these animals with three lateral lobes to the body lived. All trilobites became extinct several hundred million years ago. No other animal had, or has three lateral divisions to the body, so that the trilobite is easily recognized by school children who have had only a brief exposure to the study of earth science or geology.

When one examines a newly hatched *Limulus,* he is immediately aware of the remarkable resemblance between the young horseshoe and the trilobite. If unfamiliar with this stage in the life cycle of the crab, an individual would swear he was observing a trilobite and then become quite upset when the animal that supposedly long ago became extinct was squirming around on his microscope slide.

The so-called "biogenetic law," which says an animal during the course of its development passes through certain stages that are suggestive of its past ancestral history, was proposed by a German biologist over a hundred years ago and is an axiom of modern day biology. The biogenetic law is used to explain the presence of a tail and gill-like clefts in the throat region of the early human embryo. Interpreted as suggesting man's past evolutionary attachments, these structures are most always gone by the third month of pregnancy, but a surprising number of infants are born with tails at least partially developed (when this occurs an immediate surgical procedure is employed, and no one is the wiser). The close resemblance of the young crab and the trilobite is taken by most biologists as direct evidence the former evolved directly from the latter.

Around the turn of this century biologists devoted a great deal of time attempting to figure out which group of animals without backbones (invertebrates) gave rise to the animals, including man, with backbones (vertebrates). Many theories as to the origin of our ancestors were expounded, cursed, discussed, and discarded before conclusive evidence was unearthed which strongly suggests the group of animals that includes the sea star and the sea cucumber was the direct ancestor of the vertebrates.

The horseshoe crab or arachnid proposal was one theory enjoying considerable favor for decades. A respected scientist claimed he was able to match up certain nerves in the head of *Limulus* with the nerves coming from the brain of an extinct species of fish. This theory then was not based on the biogenetic law but on the study of comparative anatomy. The arachnid theory attracted many adherents, many of whom were well known biologists. By the 1940s, however, the sea star theory (supported largely on evidence from the biogenetic law) prevailed, and today no one even suggests a creature like the horseshoe crab gave rise to the line of evolution that led to the fish, amphibians, reptiles, birds, and mammals.

Some individuals today consider the *Limulus* theory to be capricious. Perhaps if they consider the personal habits of the expounder of this hypothesis, they will better appreciate the mental background from which the theory was conceived. There is no doubt that the originator was an unusual man. During most of his career he was stationed at the Biological Research Station at Woods Hole, Massachusetts. It is reported on good authority that on each day of the year except the very coldest he would arise at dawn, gulp down fifty cubic centimeters of two hundred proof ethyl alcohol, and then proceed to swim a half mile out into the ocean and back. After this daily routine he would retire to his laboratory where he spent the rest of the day studying the comparative anatomies of fish and crabs.

The Horseshoe Crab, A Living Fossil

Too bad for the living fossil of the river—it had a claim to everlasting fame for a few years. But, I would not be at all surprised that a million years from now will find the horseshoe crab pretty much as it is today. Not many other things on the face of the earth have the potential of accomplishing the same feat.

4

Stilettoes, Meataxes, and Bayonets

SOMETIME ABOUT three hundred million years ago a gnatlike insect succumbed to a strange new instinct that told her to try and force her mouthparts into the skin of a small lizard. Prior to this time she and her countless relatives had been fairly successful in satisfying their nutritional requirements by sucking the carbohydrate-rich juices of plants. When this ancestor of the mosquito, greenhead, stable fly, biting gnat, and other obnoxious and dangerous biting flies was finally able to penetrate the soft underbelly of the reptile and wriggle her beak into a tiny capillary, an abundance of a rich viscous material flowed into her gut, and a new physiological way of life was instigated.

A few days later the ovaries of the insect were stimulated by this new foodstuff to start the complicated process of egg development. Before this event, it had been necessary for the insect to feed several times on the sugary plant juices before ovarian activity could be detected, but one shot of the protein-rich blood of the lizard served to stimulate abundant egg production in a very short time. This gave the insect a tremendous advantage over her contemporaries who had not taken this momentous step forward and still took nourishment from plant juices.

Somewhere deep in the hereditary material (the DNA) of this creature a different combination of genetic factors had come together by chance for the first time, and the new inher-

Stilettoes, Meataxes, and Bayonets

ent instinct to bite animals was indelibly stamped into its genetic makeup. When this new combination of genes was passed on to some of the two hundred or more of the offspring, they too became blood feeders and were vastly superior to those not receiving the same dose of DNA. As this dietary practice was of obvious advantage to its possessors, the offspring that inherited the correct combination were better off than those of the same brood who were not so lucky. The eggs of those feeding on animal blood matured more rapidly and in greater numbers than those of the insects who lacked the ability to imbibe blood. Consequently, the protein feeders had a better chance of surviving and passing on the advantageous trait to a significant portion of their descendents.

These "children of blood" reproduced rapidly, passing on the genetic recipe each time to their progeny, and, within the space of a few thousand years, new species of insects evolved totally given to bloodsucking. As certain of the reptiles evolved into the warm-blooded birds and mammals, the insects experienced no difficulty in changing from the cold, unoxygenated, and inferior hemal fluid of reptiles and amphibians to the richer, oxygen-laden blood of the more complex animals. The appearance of birds and mammals turned out to be a bonanza for these insects, and they responded by developing complicated and sophisticated cutlery to mine into skin in search of the red gold. Today we see the diversity of this evolution of the drilling equipment in the forms of stilettoes, meataxes, and bayonets possessed by the species that continue their time-tested-and-proven method of obtaining nourishment for the development of their eggs.

Of all the bloodsuckers, the group of biting flies known as the mosquitoes has been the most successful from the standpoint of species numbers, numbers of individuals, and overall distribution. Worldwide the mosquitoes number approximately twenty-five hundred species with one hundred twenty having been found in America north of Mexico and about seventy-five different types having been reported from the Chesapeake Bay region. As would be expected, this diverse group of animals has developed an untold number of variations in regard to size and shape, markings, and feeding and breeding habits. These differences in mosquitoes also include the ability or inability to transmit pathogenic agents to man and other animals by injecting a drop of saliva into the puncture in the skin made at the time of the bite. This bit of salivary fluid is always introduced into the body of the victim as it contains a powerful anticoagulant and prevents the mosquito from finding itself in the embarrassing position of having its stiletto stuck fast in clotting blood. It is the saliva injected at the time of the bite which is responsible for the itching and irritation we may experience.

Of all the sundry varieties of mosquitoes plaguing mankind, none are more ferocious, prolific, and downright bloodthirsty than the so-called salt-marsh variety which is worldwide in its distribution. Actually the name salt-marsh mosquito encompasses two or three distinct species, but each of them is just as pestiferous as the others. The female salt-marsher secures her blood meal from a human, muskrat, deer, or bird and quickly matures her eggs. These embryonic demons of the marsh are deposited in a damp depression that the mother knows by instinct will be periodically inundated for short periods of time. The young mosquito wriggler or larva matures inside the tough eggshell, and when the depression is flooded, it uses an ingenious can opener type device on the top of the head to cut its way free from the shell. After a period of ten days to two weeks (depending on the temperature) development in the water, it metamorphoses into the flying adult with which we are so familiar.

As with many other types of animals the female is the more dangerous sex of the species. The male lives only to mate with the female and deposit untold thousands of tiny sperm cells into a special reservoir adjacent to her reproductive tract. There the male cells will remain viable until they are released by the female—one by one—to fertilize the eggs as they pass Indian file down her oviduct to the outside. Having fulfilled his biological duty, the male has no other reason to continue his existence, and so he drops to the bed of the marsh and,

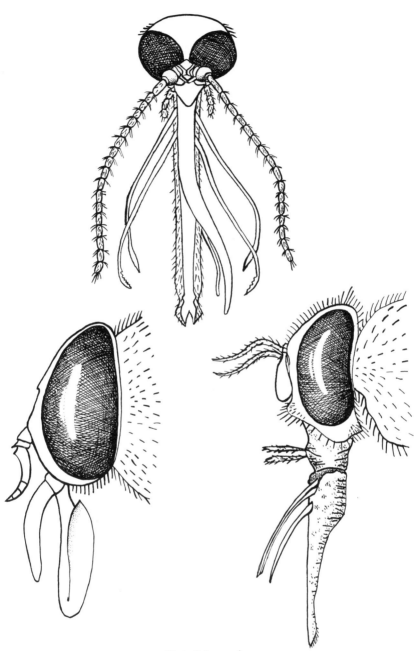

Top: Mosquito;
bottom left: Greenhead; *bottom right:* Stable fly

without fanfare, expires. During his brief lifetime he has had no reason to suck blood, but he may have acquired a few calories from the nectar of flowers to supply the energy required to accomplish the reproductive act. It should be noted, however, that Ed Bedloe, a longtime resident of Nanticoke, believes the male mosquito has another function in life. This observer of the natural history of mosquitoes was aware that the male did not take blood, but he was adamant in stating the male had the duty of guiding the female to her feeding grounds during the evening hours.

Shortly after she is inseminated, the female has but one goal in life—the procurement of blood to nourish her developing eggs. A human being presents a delightful expanse of relatively bare flesh for the insect to attack. It is more difficult to probe through countless layers of feathers and hair before striking blood, but if a human being is not available the female will persist in her efforts to penetrate another animal. The feet, around the eyes, and beak or nose region are relatively unprotected by the shields of hair and feathers and are favorite dining spots. The seminude young of birds and mammals are also a favorite target, and great numbers of fledglings are undoubtedly killed at certain times of the year by exsanguination.

When the climatic conditions are just right, untold legions of mosquitoes emerge from their marshy incubator at about the same time. I have seen swarms arising from the wetlands of the Nanticoke in such numbers that a visible cloud is formed, and, if you are not sharpsighted, you would think the marsh was afire. I have observed similar swarms of mosquitoes on only one other occasion in my travels to various parts of the world. During the summer of 1951 when I was in the U. S. government's oil exploration camp at Umiat, Alaska located some one hundred and fifty miles south of Barrow, the mosquitoes emerging from the wet tundra were so numerous that they would blot out the olive drab color of a wool army service cap exposed to their landings for thirty seconds.

Salt-marshers have the nasty habit of not remaining in the area from which they hatch and range far and wide in search of their necessary protein. Residents of nearby towns and villages as well as the numerous cottage owners are favorite victims. Entomological studies on the flight range of these insects have shown without question they will migrate several miles from their breeding area. One such research project, in which the bugs were marked with radioactive tracers, revealed that some members of a brood may travel as far as forty miles from home in their sanguinary quest.

It is during this phase of the mosquito's life cycle that life for man and beast can become almost unbearable. "When the skeeters are on the wing," as the local residents say, one might as well forget any plans for any extended outdoor activity. The Indians inhabiting the region of the Nanticoke hundreds of years ago learned there was no way to cope successfully with the problem during the months when the bugs were at their greatest abundance. They routinely left the area during the summer to take temporary residence on the windblown shores of the Atlantic some thirty miles away.

While at home on the banks of the river, prior to the annual trek to the east, these Indians attempted to achieve some protection by covering the exposed parts of their bodies with rancid bear grease or the rendered fats of other animals like the opossum and raccoon. The body had to be entirely covered with the material as any chink in this organic armor was quickly discovered and exploited by the bloodthirsty mosquitoes. If one tiny patch of skin was left unprotected, numerous females would move in and scramble for a place at the feeding trough. The Indians would also try to protect their campsites—especially in the evening—by lighting smudge pots of green pine needles or shats, as most people on the Eastern Shore call them. The pungent smoke emitted was supposed to kill or repel the insects but it is probable the lungs of the Indians were more adversely affected than the simple breathing pores and respiratory tubules of the salt-marshers.

Contemporary residents of the Nanticoke have resorted to concoctions just as odiferous and foul smelling as rancid animal fat in attempting to protect their bodies from the fiery bites. And, a modern day version of a mechanized smudgepot is employed by numerous state employees who dash in and out of the inhabited areas with their fog and smoke machines attached to four-wheel drive vehicles. It is extremely doubtful the insecticide-laden smoke and mist that belches forth from these infernal machines have any more effect on the mosquitoes than the primitive smoke generators employed by the Indians did.

Watermen tending their crabpots during the summer have been known to smear their bodies with grease and oil from the engine and boom of their workboats. Simply being out on the water a few hundred yards from the marsh offers no immunity from attack. Working under the hot summer sun causes perspiration to pour forth in copious amounts. The ever present breeze will pick up the aroma of human sweat and transport it over a wide area. To the vampires resting in the marsh, this infinitesimal chemical aroma is quickly detected by the special olfactory organs on their antennae or feelers. The response to this dinner bell is similar to the reaction of a company of firemen leaping into action when the telegraphic alarm is sounded. With utmost haste they will proceed directly to the site from which the signal originated.

There is nothing more frustrating and nerve-racking than to be engaged in some sort of activity that requires manual dexterity, and, just as the operation is in a critical stage, be abruptly distracted by the burning sensation that marks the entrance of the mosquito's proboscis. Once my outboard stalled while slowly running up a gut known as Jack's Creek. As the motor had been missing and discharging a smoky exhaust, I reasoned I had mixed in an excessive amount of oil with the gasoline, and the spark plugs were fouled. Simple matter to correct—remove plugs, wash plugs in gasoline, dry them, reinsert into the block of the motor. Everything was proceeding according to plan until it became necessary to unscrew tediously the last few threads on the base of the plug. I had accomplished this, and with the greatest of care was easing the part out of the block when—Wham! A redhot needle was jammed into my neck. For an instant all thoughts of the spark plug were erased from my mind as I responded auto-

Stilettoes, Meataxes, and Bayonets

matically to the stimulus by swatting at my tormentor. Plunk!, the plug dropped from my hand into five feet of water and quickly disappeared into the oozy, muddy bottom of the gut.

Of course, I had no extra plug, and, as luck would have it, no other boats were in sight that might have answered my plea for a tow. I had no alternative but to pick up a flimsy canoe paddle and lay to it. The two miles or so back across the river to the cottage required about three hours to negotiate as the tide and wind were against me. Eventually I propelled the boat back to the anchorage in front of the house and fell exhausted on the sand of the beach. Every muscle in my body screamed out in agony, and my face was red and puffy from the hundreds (it seems) of mosquito bites I had taken during the paddling fiasco. I laid there cursing the mosquitoes, myself, and the world in general and took a solemn vow never to venture forth on the river again without an extra spark plug.

Some old-timers along the river still adhere to mosquito abatement procedures that have been handed down for countless generations. The fragrant grass, citronella, imported from southern Asia, has been used as an insectifuge for as long as many inhabitants of Nanticoke, Bivalve, Tyaskin, and Wetipquin can remember. Its active ingredient is both repellent and toxic to insects when incorporated into a candle and changed to a thermal aerosol when the taper is burned. The only fly in the ointment is that citronella is also toxic to humans when it is suspended in smoke. A family might acquire some relief from the mosquito problem in their house or yard if citronella is burned, but they may suffer neurological disturbances if too much of the poisonous smoke is inhaled. Dizziness, lack of coordination, blurred vision and speech, as well as severe, persisting headaches are commonly reported side effects of overexposure.

The burning of punk to create a smoke-screen obnoxious to the airborne vipers is also a common practice. Punk is merely a piece of rotten wood carved from an old stump and ignited. If the decayed wood is in the proper condition, it will smolder for hours and do almost as good a job of dispersing the pests as citronella. In addition, no neurological difficulties are to be expected from the inhalation of the smoke from punk, and it is cheaper and easier to procure as well.

The extracted oil of citronella is frequently combined with a light mineral base oil and used as a skin repellent. In fact, oil of citronella was the most widely used insect repellent in the world prior to World War II, and it does afford a considerable degree of protection against mosquito bites of all kinds. It is possible to buy oil of citronella in certain parts of the United States today, but the much more effective synthetic chemicals developed during the last twenty-five years have virtually eliminated a demand for it. A year or so ago I noticed two bottles of citronella on a shelf in the small general store in the hamlet of Bivalve. The proprietor of the establishment stated the two bottles came with the store when he purchased it some ten years previously, and he had no idea how long they had actually been on the shelf. It must have been a long time as the labels were so faded one could hardly decipher the printing. The storekeeper allowed as a quarter ought to be enough for one of the bottles, and I quickly purchased one as a souvenir of another era.

It is difficult to believe, but a retired waterman, also from the same village of Bivalve, once told me he had previously kept a bottle of stale human urine in the cabin of his boat. When the mosquitoes attacked in force, he had no qualms about applying the material to his face and other exposed parts of his body. He claimed this nitrogenous product gave him complete protection for several hours. The gentleman had undoubtedly given a great deal of thought and experimentation to the various and sundry uses to which urine could be put, because he also suggested to me that the best way to relieve the blazing pain of a sea nettle sting was to rub the affected area with freshly produced urine. The fermented product was not nearly as effective as the liquid just passed from the bladder.

Oils of various types are recorded in history as having been utilized at one time or another as insect repellents. Among those which have been suggested are anise, bergamot, cassia, cedar, clover, eucalyptus, lavender, citronella, pennyroyal, tar, and turpentine. Also advised have been such substances as camphor, castor oil, creosote, crushed pepper, eau de cologne, iodoform, kerosene, the juice of leeks, lemons and limes, menthol, musk, naphthalene, nutmeg, pyrethrum, and vinegar. The fact that so many chemicals have been tried suggests that none has proven satisfactory.

As far as I can determine the residents of the Nanticoke have routinely employed only two of the above substances in preparing homemade repellents—citronella and pennyroyal. Various residents of our area have been persuaded to share with me the family recipe for a mosquito repellent, and I shall share them with my readers for possible formulation and use.

One woman told me her mother used to mix three ounces of pine tar, three ounces of petrolatum, and one ounce of citronella in a can placed in hot water. When the ingredients had mingled completely the concoction was "put up" in small widemouthed jars (such as empty petroleum jelly containers). The viscous nature of the petrolatum served to keep the concoction on the skin longer than a more fluid mixture.

Another acquaintance recounts the formula for a supposedly effective repellent mixture that was frequently used in the good old days. Pine tar, olive oil, and an ounce of pennyroyal were the basic ingredients. Pennyroyal was supposed to be superior to citronella as the latter material was said to have been responsible for various skin conditions such as hives, eczema, and psoriasis. This individual suggested that if the mosquitoes were especially numerous and fierce, a shot of camphor would serve to enhance the effectiveness of the repellent. It was quickly pointed out, however, that if camphor crystals were utilized they should be dissolved first in good moonshine. Good moonshine presumably refers to a distillate in the neighborhood of 190 proof, or almost one hundred percent grain alcohol.

Biologists believe each species of life on the face of the earth evolved to occupy a specific ecological niche, and all of these different environments are interrelated along with their inhabitants. Disturb the environment or the occupants of one and the others will be upset in a chainlike reaction. In other words all species are dependent to a greater or lesser degree on each other, and each type has a contribution to make to the entire ecosystem.

I have often wondered how important the mosquitoes are in the overall picture. In the aquatic stage they serve to implement the diet of larger organisms, and when they have assumed their flying status, they also make up part of the daily bread of insectivorous birds, bats, and even other insects. But, the damage they do to species of wildlife, such as destroying many of the young and transmitting pathogenic agents of one kind or another to man and other animals, seems to far outweigh their questionable value as part of the overall food chain.

On the other hand, I was given a different slant on the importance of mosquitoes and other bloodsucking insects while engaged in conversation with a wharf-habitué in Tyaskin. We were discussing the current supply of crabs in the river, and the salt-marshers were attacking with a vengeance. Though I had repellent on my exposed skin, many of the monsters were biting through the T-shirt I was wearing. My acquaintance was not wearing any sort of repellent and was being fed on at a terrible rate. At one time I counted five blooded females resting on the lobe of one of his large ears. As I attempted to terminate the conversation, giving the abundance of mosquitoes as the primary reason, he delayed me. "Let 'em bite on you for a while, it'll do you good." As my nerves were already frayed to the snapping point, I rudely retorted, "Since you are so God-damned smart, I would like you to explain how in the hell it is going to be good for me to stand here and have these insects suck my blood?" He came back at me with just as good as he had received and replied, "I didn't realize you were so damn ignorant that you didn't know all livin' things is always makin' blood in their bodies. Iffen we didn't have the skeeters and greenheads to suck some of it out now and then we all would soon blow up and bust!"

Another *possible* reason why mosquitoes were placed on earth by the powers of nature came to my attention during my aforementioned trip to Alaska in 1951. Part of our entomolog-

Stilettoes, Meataxes, and Bayonets

ical work was conducted at the village of Umiat, and it was during my week or so stay there that I made an observation which may explain more satisfactorily the reason for the mosquito's presence on earth. Umiat is located on the Colville River in the tundra of north central Alaska. Numerous small streams feed into this large river. I was delighted to observe that these crystal clear feeders were loaded with arctic grayling and trout. One evening I ventured into the tundra to try my luck, and you can be sure I was well covered by protective clothing, repellent, and a headnet. As I looked down into the eight feet of water that appeared to be no deeper than three, I could clearly see numerous grayling milling about the bottom. The artic grayling is a beautiful fish, but it has the reputation of frequently not cooperating with the angler. I offered them every fly and nymph in my repertoire, but none of them showed even the slightest interest. It was a maddening and frustrating experience to have grayling so near, yet so far away, and I was just on the verge of returning to camp to pick up a hand grenade with which to blast them from the icy water. Just then I happened to notice a grayling rise to the surface and suck in a large mosquito that was resting there. As I continued to watch, I saw the same thing happen again and again. Fortunately, I had a bottle of clear nail polish in my kit (to repair tackle and to stop the itching of a mosquito bite when nothing else seems to do the job). Easily I captured several of the large insects and, with the nail polish, affixed one to a small number-twelve hook attached to a fine leader. I drifted this bait down the stream and almost immediately a half-pounder rose and gulped it down. I took six others that night in the same way, and the fights they put up in that cold water were a true test of my ability with the fly rod. I must also report they were about the best tasting fish I have ever had the pleasure of eating. But, the point is, I think I may have stumbled on to the answer of one of the great mysteries of ecology—why mosquitoes are present on earth. Undoubtedly, they are here so men may use them to take arctic grayling at Umiat! In any event, this theory makes more sense to me than the "Blow Up and Bust" hypothesis.

Along the Nanticoke problems with mosquitoes may start as early as the first of April and may extend well into the fall. I have experienced difficulty with them as late as Thanksgiving Day. In addition, sometime around the last part of June, and lasting into early September a robust, determined, and ferocious bloodsucker makes its appearance in the area and adds untold misery to the already mosquito-weary residents. This avid and determined bloodsucker is the salt-marsh greenhead fly, a member of the family of flies that includes the horse, deer, and *Chrysops* varieties.

Where the mosquito has a long, slender, stilettolike implement to pierce the skin, the greenhead is possessed of a wedge-shaped meatax. When this monster tears into your skin, you know you have been bitten. Many times the skinny proboscis of the mosquito will not make contact with a nerve ending in the skin, and it is only when she has fed to repletion and flown away and the irritation of the saliva begins that you know you have been bitten. Not so with the greenhead! There is no way her wicked weapon can enter the skin without severely antagonizing numerous nerves, and sometimes the pain of the bite borders on the excruciating. The anticoagulant saliva of this fly is more potent an irritant than that of her distant relative, the mosquito. Reactions to the bite vary with the individuals, but, in those that happen to be especially sensitive, the bite site may swell and become inflamed over a one inch or more area. There is no way of determining how many times I have been bitten by mosquitoes, but I can say that I cannot remember a bite that resulted in bleeding. Many times, however, the bite of the greenhead has produced an actual flow of blood from the ragged wound made by her terrible mouthparts. Many of the residents of the area are actually afraid to participate in outdoor work or play when these devils are numerous.

About the only thing one can do to protect himself from their bites is to cover every bit of the body with thick clothing, including a net over the head. T-shirts or tight fitting garments merely slow them down a bit. Of course, the hot muggy days of July and August are certainly not conducive to this sort

Stilettoes, Meataxes, and Bayonets

of attire. The standard insect repellents containing ethyl hexanediol and diethyl toluamide (Deet), that work well against mosquitoes when used as directed, have very little effect on greenheads. Greenheads that are determined to bite will quickly break through a repellent barrier that is effective for lesser bugs.

Once a mosquito is swatted it is almost always dispatched on a one-way trip to the insect's happy hunting grounds. A greenhead requires a resounding whack even to deter its biting activity. Many a time I have swatted a greenhead on my body and gotten in a good lick too. But, more often than not the stunned fly will drop to the ground, or deck of the boat, shake its head as if to clear the cobwebs, and then fly straight back to me and attempt to resume its interrupted meal.

One of my friends in Bivalve says the way to "get 'em" is to wait until they "pitch" before you slam away. By "pitch" he is referring to the behavior of a duck or goose coming in to land on the water. Just before contact with the surface, the waterfowl pitches its feet and legs forward to make impact and spreads the wings in a braking maneuver. Any duck or goose hunter knows that the easiest shot he can have is when the bird is pitching, and so my friend reasons the best time to take a shot at a greenhead is when it is approaching the body in preparation for landing. Unfortunately, slow motion pictures of greenheads (as well as mosquitoes and other flies) in the act of touching down do not reveal any of the behavior patterns of ducks and geese that make the latter easy targets.

The greenhead is such a tough and vigorous insect that chemical measures employed for mosquito control have little or no effect on him. Generally speaking, the larger the insect, the greater the amount of insecticide required to kill it. A large female greenhead may measure almost a half an inch and weigh fifty times more than a salt-marsh mosquito, and so a good bit more of the toxicant is required. With the rates of application of chemicals for mosquito control under question as to their effect on nontarget organisms, there is no way the dosage rates can be increased to the point where greenheads are killed without seriously damaging other members of the fauna.

Many times during the summer I have been plagued at my cottage with both varieties of bloodsucker and welcomed the arrival of the four-wheel drive pickup from the government's abatement agency with the mistblower on the back. The truck would make a quick circle in the yard as a fine cloud of insecticide was blown hither and yon. Almost immediately the mosquitoes would be affected and their nuisance quotient reduced to almost zero (they were either killed or repelled from the yard). But, the greenheads would seem simply to take a whiff of the odiferous pesticide, sneeze, and resume their attack with renewed vigor. I have aimed pressurized insecticide bombs directly on greenheads at a range of only a few inches, and never have I observed the typical twitching, loss of nerve coordination, and convulsions that are characteristic of other insects that have received a lethal dose. Many of these bug bombs that are so popular today utilize freon gas as a propellent for the toxic ingredients. Freon is a refrigerant, and when the fly is hit with a shot of this numbing gas, it becomes quite sluggish for a minute or so. Now is the time to crush the life out of the demon with your hand or swatter. Normally it is next to impossible to do this because of its acute vision.

Instead of having a normal size head with two eyes, this insect has two eyes and a very small head. In fact, most of the head region of the body is made up of the two enormous green eyes—hence the name greenhead. Its vision is normally so discerning that the slightest movement of a hand or other tool of destruction in its direction is immediately detected, and the bug is able to react instantaneously and dart away to avoid the blow.

Greenheads utilize the marsh for breeding as do the mosquitoes, but the young or larvae do not require free water in which to develop. Larvae develop in the top layer of the muck of the marsh just above the high tide level. As many as one hundred of these footless and wingless maggots may be found in a square yard of marshland. The maggots forage through the wet thatch and decaying grass in search of invertebrate animals such as snails and other insects on which to feed.

Stilettoes, Meataxes, and Bayonets

When the supply of these prey species runs low, the greenhead larvae have no qualms about turning on their own kind and practicing the ancient rite of cannibalism. This despicable habit (from the point of view of humans) insures that only the strongest shall survive to mate, feed, and perpetuate the species.

When cold weather arrives in the fall the larvae burrow a few inches into the muck of the marsh and enter a state of physiological inactivity. With the advent of spring the life processes are reactivated, and the larvae move to the surface and feed for the first time in many months. This nourishment triggers their entrance into the pupal or resting stage during which the anatomy and physiology of the insect are drastically altered. The wingless, wormlike, nonblood-feeding larvae with chewing type mouthparts are transformed into the familiar greenheads with the meatax type mouthparts and the insatiable appetite for blood. The adult emerges from the pupal cocoon in late spring. Adult flies mate a few days after emerging, and, unlike the mosquitoes, the female will deposit her first egg mass in the marsh without having previously taken a protein-rich blood meal. Apparently enough protein, acquired by the predaceous larvae, has been held in reserve to mature the first batch of eggs, but only the first batch. Subsequent clutches of eggs require the all too familiar blood meal to be taken from wildlife, livestock, or humans.

As with the mosquitoes, only the females take blood—the males resorting to sapsucking to get the energy they require. To find the source of reproductive power the older females migrate out of the marsh to the fields and headlands where they await their victims. Greenheads are long-lived and have been known to survive for three or four weeks. As a result of this longevity, the population continues to grow as the summer progresses, and by the middle of August their numbers will have reached staggering proportions. This time of the year along the river is known as "The Fly Season" and one can appreciate the aptness of the designation.

E. P. Catts and Elton J. Hansens, professors of entomology at the University of Delaware and Rutgers University respectively, have devised a simple trap for use during the fly season. This device, though not capable of eradicating the greenhead, will reduce their numbers. This plywood boxlike snare is some thirty-two by thirty-two inches wide and twenty-four inches tall. It stands on legs set two feet off the ground, and the bottom and top are covered with household screening. The bottom screen is folded back in such a manner that an entry portal is created. The sides and legs are painted a glossy black to contrast with the landscape and to capture solar heat. Greenheads are attracted to glistening, warm targets, and no bait is used. For best results the trap is placed on the rim of the marsh or along the open edge of wooded or shrubby areas. Professors Catts and Hansens report that a capture rate of three thousand flies per day is not uncommon. During testing in 1972 forty of these traps collected a total of one hundred eighty-six thousand greenheads.

    I constructed one of these traps for the backyard and placed it next to the marshy area bordering our property on the north. About the best day I had with the apparatus was early in August when I captured three hundred flies in one twenty-four hour period. I cannot really say this catch made any noticeable dent in the population in my backyard, but I did have the satisfaction of knowing that three hundred of the devils would not be around to plague me or propagate their kind.

    I reported this accomplishment to one of my neighbors who occupies his cottage more or less on a year round basis. He replied, "So what? So you trapped two or three thousand all year. What's the big deal? When you are dealing with millions of the bastards, your catch is not even a drop in the bucket!" Of course, he was accurate in his assessment of my efforts to help mankind, but if every property owner could be persuaded to employ one or two traps on his grounds (a unit costs next to nothing in materials), the numbers collected would be more than a drop in the bucket. Within a few years this destruction of the breeding stock, albeit ever so gradually, would eventually be reflected in a significant reduction in the overall population.

Stilettoes, Meataxes, and Bayonets

The use of one organism to control another is known as biological control, and dramatic achievements have been accomplished with this method in the fields of economic and medical entomology. In most cases, however, the use of biological controls is not satisfactory when used alone as they must be supplemented with chemicals and physical procedures. Many residents of the Nanticoke region have attempted to alleviate their biting fly problems with the use of the purple martin as the biological control agent. Long before the streamlined, mass produced houses for these large swallows were placed on the market in the mid-sixties, local inhabitants encouraged the nesting of purple martins by erecting all sorts of houses in the hope they would set up housekeeping in their yards. The Indians of the area recognized their value and offered large gourds with an entrance hole cut out as nesting sites. Every now and then, while cruising the countryside, one will see the gourd-house type of bird dwelling in someone's backyard.

Purple martins are insectivorous birds that will devour any insect they encounter in flight. Certainly, large numbers of mosquitoes, greenheads, and stable flies are devoured, but the martins do not differentiate between harmful and useful insects. Studies have revealed the birds eat just as many useful species as detrimental types. During the mid-sixties (after the widely acclaimed *Silent Spring*) martin houses were pushed on concerned consumers with the exaggerated claims that a purple martin would ingest *forty thousand* mosquitoes a day. This was the nonchemical, ecologically compatible way to solve the problem. In fact, the small town of Griggsville in west central Illinois adopted the official name of the Purple Martin Capitol of the World. Thousands of aluminum houses were set up in the town and the city fathers boasted far and wide that the place was entirely free of mosquitoes. It should be noted that one of the industries of that town was the manufacturing of martin houses.

Unfortunately, many people believed these tales of mosquito-free environments and rushed out to purchase untold thousands of these houses which were usually sold with a special telescoping pole that permitted the lowering of the residence for periodic cleaning. Certainly martins will help alleviate the fly problem in much the same way as will the specialized trap for greenheads, but these control measures alone are not enough to produce any noticeable reduction in numbers.

As one drives through the village on the eastern shore of the Nanticoke he will notice martin houses (usually of the deluxe aluminum variety) adorning the yards of many of the residents. When you talk with the people who have erected the houses in which martins have established residence you get a variety of answers to the question of how effective are the birds. Some people swear by them while others freely admit they can notice no reduction in insect numbers. I suspect the individuals in the latter group are more truthful than those who claim the birds have rid their premises of the despised pests.

Lt. Col. Manfrid Ernst (USA-Ret), a longtime resident of Nanticoke and astute observer of natural history, believes the martin houses and their residents accomplish more psychological than biological control. After all, if it is generally believed that one martin will destroy forty thousand insects a day, the people translate this belief into a "mosquito-proof" yard because they have a birdhouse.

I have yet to meet a man who expressed any difficulty whatsoever in immediately recognizing a mosquito or greenhead, but the same cannot be said when it comes to another type of bloodthirsty denizen that is all too common along the Nanticoke. The stable fly is an overgrown gnat which bears a striking resemblance to the common everyday housefly. In reality the housefly is a little larger than the stable fly and the coloration is slightly different. It takes a well trained eye, however, to pick out these differences without the use of at least a hand lens.

The similarity between the two insects ends when the mouthparts are examined. The housefly has sponging and lapping mouthparts for the intake of nutrients which have previously been partially digested by the regurgitated contents

Stilettoes, Meataxes, and Bayonets

of its stomach. The stable fly has a piercing and sucking, bayonetlike proboscis that protrudes forward in a menacing manner. Frequently, the stable fly is referred to as "the biting housefly," and many people wonder how and why a harmless housefly can quickly convert from a diet of liquified organic matter to blood in the twinkling of an eye. Many have expressed beliefs that an assortment of factors (usually associated with climatic changes) are responsible for the draculalike change. Of course, none of these is true, for we are dealing with two distinct species of insects which are not very closely related. The superficial similarity between the two is sometimes called parallel evolution—where two relatively unrelated organisms adapt to the same environmental conditions in a similar manner and thereby grow to resemble each other. The similarities between a shark and a porpoise are a good example of how evolutionary forces operate to create lookalike species that are constructed to fit their particular ecological environs.

The stable fly (or dog fly as it is sometimes called) is a vicious biter. It draws blood quickly and fills to capacity in from three to four minutes if undisturbed. However, ordinarily even when undisturbed, it changes position frequently or flits to another host to continue its meal. This nervous habit results in several skin punctures made by the same fly. Unfortunately for most warm-blooded animals, both males and females are bloodsuckers and will travel many miles in search of a conquest.

The stable fly is not a product of the marshes but of the beaches where it breeds in decomposing deposits of washed up grasses that mark the high tide level. The name stable fly has long been used because this insect frequently breeds in decaying straw and hay found around stables, barns, and pastures.

Maximum populations of these flies more or less coincide with the explosion of greenheads in midsummer, so the term "fly season" is well taken. Like the greenheads they are robust insects that are difficult to kill or repel. The dosages of chemicals applied for the control of adult mosquitoes offer little relief from these menacing creatures. The skin repellents work a little better than with greenheads, but none I have ever used turned out to be effective for other than a very short time.

These flies are crafty and ingenious, and they seem to have at their disposal a bag of tricks with which they can outwit the supposedly smartest animal on earth—man. Frequently I have found it necessary to haul up the anchor from a fishing spot near the beach and retreat to the open water to escape the stable flies that had congregated aboard. Usually this retreat has been made with a full throttle in an attempt to blow away the flies and leave them behind. After slowing down or anchoring at another spot, hordes of the insects would immediately begin the attack anew. They had sought refuge from the wind in the protected recesses of the boat and were ready to swing back into action as soon as conditions permitted. For some reason these little vampires like to bite the feet and toes, and this habit can result in considerable unpleasantness for the barefooted victim who may be attempting to steer a boat and swat flies at the same time. Again, like the greenhead, a forceful whack is necessary to dispatch the stable fly as a lesser blow will only benumb him for a short time. Before you know it he will be back at his dirty work.

It is interesting to note that at the beginning of this century there appeared a scientific report that offered proof the stable fly was the vector of the dread virus of poliomyelitis. Subsequent experiments in 1913, however, proved there was no basis for the previous claim. Medical entomologists suspect any bloodsucking insect has the ability to transmit certain disease organisms, but so far they have been unable to prove that either the greenhead or the stable fly are carriers. Of course the roles of various mosquitoes in the epidemiology of many human and animal diseases are well known, but that is a story for another time.

There are other bloodsuckers to cope with along the Nanticoke including ticks, chiggers, deerflies, punkies, no-see-ums, and the legendary blue-tail fly, but these pests are benign in comparison to the ones we have discussed. Taking everything into consideration, there are times when this "Land of

Stilettoes, Meataxes, and Bayonets

Pleasant Living" can be *not* so pleasant. But, this is part of the overall picture of the natural history of the area. Upon reflection, I have arrived at the conclusion I would prefer to have it this way. The pestiferous bugs serve to hold down the development of the region to a certain extent. Many have been the thin-skinned city dwellers or suburbanites from Baltimore, Philadelphia, or the hub of the Eastern Shore, Salisbury, who have toyed with the idea of acquiring a bit of land with a cottage "up on the river." More often than not when these individuals finally decide to do something about purchasing or building their dream cottage and seriously make on site inspections of available real estate, they beat a hasty retreat to the sanctuary of the city where the only stilettoes, meataxes, and bayonets they may be exposed to are wielded by fellow human beings.

Come to think of it, this deterrence of nature's to the invasion of the land by those who do not understand it could be a reason why the bloodsuckers of the Nanticoke were placed on earth. This, indeed, may be the true answer to the perplexing question of "Why?"

5

Garfish Marauders

TWO DAYS HAD PASSED since the Eastern Shore had been inundated with four inches of rain, and the effects of the runoff were evident in the lower parts of the river. Tides were running a foot above normal, and a large amount of flotsam was drifting by the cottage. Boaters were smart if they kept a sharp lookout for driftwood or partially submerged logs.

The normal salinity of the southern part of the Nanticoke at the beginning of summer is about thirteen parts salt to one thousand parts water, and the river has a definite saline taste. I took a sip of the Nanticoke and only a faint taste of salt could be detected. The insurge of fresh water upstream had diluted the salt concentration down to about one or two parts per thousand.

After determining the salt content I looked out to my nylon gill net stretched between two poles embedded in the bottom and had a pretty good idea of what I would find in it. Sure enough, as I waded out to the net I could see the plastic floats on the top support line moving abnormally in the current. This indicated one or more large fish had become entangled in the meshes. Three large gars (garfish) had become entangled and were doing their best to gain their freedom. In the process, they were successfully destroying large sections of the net. Each of the fish, weighing about four pounds, had been caught in the nylon net by the thousands of sharp teeth that line its elongated jaws.

Garfish Marauders

With considerable difficulty I extracted them from the net and quickly dispatched them to storage in the freezer. The net was severely damaged, and it was doubtful it could be repaired. In any event I knew it had to be removed from the river as the garfish I had caught were merely the advance guard of a larger number heading downstream into the semi-fresh water on a periodical foray. Gars are not normally found in this section of the river where it is usually too salty for them, but when the salt content drops below five parts per thousand, they migrate out of their fresher natural habitats and follow the fresh water downstream. In some years when there is excessive runoff they may work their way out of the tributaries and into the Bay itself for relatively short periods.

As my gill net was only one hundred feet in length and covered less river than when it is set out in a bowed fashion, the catch of three garfish indicated these marauders were on the move and coming downstream in considerable numbers. The river in front of the cottage is a little over a mile wide, and if a hundred foot section had yielded three fish, one knew there were plenty more of them out there in the remainder of the expanse. (Incidentally, Maryland law permits a property owner to utilize a net up to one hundred yards in length in front of his property after payment of a small fee.)

I retrieved the net and suspended it between a group of pine trees to dry. There I could fully inspect the damage and decide to repair or replace it. A superficial examination indicated there was little chance the net could be repaired, but I gave this matter little thought as I had plans for a more sporting activity that comes along only now and then—the taking of garfish on rod and reel.

I grabbed a six-foot spinning rod and a tackle box I was sure contained some special equipment that would be necessary for this expedition, and, after seining a bucket full of small Norfolk spot and hardheads, set out in quest of garfish in my small outboard-powered pram. As the tide was ebbing, I decided to drift along a shallow stretch between Hatcrown Point at the entrance to Tyaskin Harbor south to the jetty marking the entrance to Jackson or Bivalve Harbor. I hooked

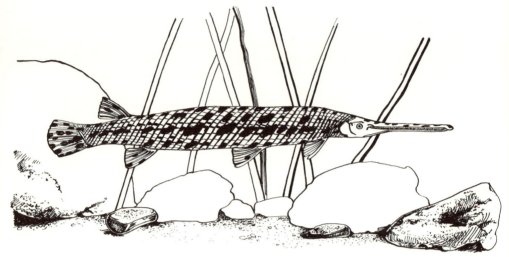

Longnose garfish

Garfish Marauders

a small spot onto the terminal tackle rig and suspended it about two inches beneath a large red and white plastic float. I flipped this rig in front of the boat and let it slowly drift downstream. The rod was placed in a holder, and I sat back to steer with a paddle while all the time keeping a close eye on the float.

The boat had not meandered far downstream when the bobber started moving off at an angle. After what seemed like an eternity (actually only a few seconds) the float disappeared, and I knew it was now or never. I brought the stiff rod tip up and backwards with all the strength I could muster, and I knew immediately that success was mine. The fish on the other end of the line shot off downstream freely stripping nylon from the reel against the buzzing of the drag. After fifty feet or so of monofilament had been ripped off, I put the brakes on lightly in an attempt to turn him back upstream. This pressure goaded him to explode clear of the water and do a brief little dance with his tail on the surface.

The fish fought as if he were possessed with some ichthyological demon, but I was able to keep up with him and maintain constant tension on the fifteen-pound line. After a couple of minutes I felt the ebbing of his fighting spirit and was gradually able to regain some of the line I had lost when he made his initial bid for freedom. The fish was not ready to quit, however, and made several desperate lunges into the depths as he neared the boat. In about five minutes he was exhausted, and I was able to bring him alongside. With the greatest of care I slipped my hand under his belly and, with a quick flip, maneuvered him into the boat. As he hit the gunwale his combative instinct revived, and it was touch and go for a time as to who would jump out of the boat first—the fish or I. Eventually he calmed down a bit, and I was able to administer the coup de grace behind his eyes with a miniature baseball batlike implement frequently referred to as a priest.

I had taken and managed to subdue a two-foot longnose garfish weighing at least seven or eight pounds. During the next two hours or so I was able to connect with about a dozen more and boat six or seven. These fish also eventually ended up in the freezer to be utilized at a later date as bait for the crab pots or transformed into a smoked gastronomical delicacy.

Garfish (sometimes called garpike, pike, or billfish in various parts of the country) are universally hated by the residents of the Nanticoke. When the gars venture downstream from their normal hunting grounds at the headwaters of the river, as previously indicated, they frequently do considerable damage to the nets of both amateur and professional fishermen alike. Gars are also believed to decimate the populations of rockfish, trout, spot, hardheads, and bluefish, and they are blamed for poor catches of these "desirable" fish at certain times of the year. To insult further the human ego, they are all but impossible to catch on conventional fishing rigs. And, besides, as the natives reason, "Who in the hell would want to catch one anyway?" The natives consider them trash fish only a starving man would attempt to eat.

Boys of all ages can sometimes be observed shooting at garfish with twenty-two caliber rifles from a bridge spanning Wetipquin Creek as it flows into the Nanticoke via Tyaskin Harbor. Frequently outboard boaters try to chop them to bits with their propellers when the fish swim lazily near the surface. Sometimes gars will attack live fish on a stringer or chain hung over the side of a boat and do considerable damage to the catch before they can be detected. Of course, this habit does nothing to endear gars to the local anglers.

Bill Robinson, a disabled waterman who fishes from the bridge at Wetipquin each day of the year that weather permits, has a way of dealing with any garfish he may occasionally snag. He props the long snout open with a stick and then releases the unfortunate fish to swim crazily in circles until it drowns. Robinson fishes every day to put food on the table of his large family. He keeps almost anything he catches—including small catfish, perch, and spot in the fingerling class. Jokingly he says it takes a bushel basketful of fish a day to feed his tribe, and, after visiting his home and making a headcount, I see what he means. Robinson considers the numerous eels he catches delicacies but will not grace his table with a gar even though I have pointed out to him that gars are "clean" fish, feeding only on

Garfish Marauders

other live fish. He thinks of them as aquatic garbage collectors and will never be convinced they are anything else.

There is something about the grotesque structure of garfish which instills in men both fear and hate. Garfish do not look like most other fish. Furthermore, the hundreds of needlelike teeth lining the bony, elongated jaws of garfish contribute considerably to these human emotions. Robinson tells me he is very careful when handling a gar as a few years ago a friend of his was attempting to do the stick-in-the-jaws trick, and the gar clamped down on his hand. According to Robinson, the fish had to be decapitated with a sharp ax before the man's hand could be removed. The wounds made by the teeth refused to heal, and the man eventually had to have two fingers amputated. Robinson does not go so far as to state positively that the bite of a garfish is venomous, but the implication is strong that some sort of destructive chemical substance is present in the fish's saliva. However, according to ichthyologists there is absolutely no basis for this supposition.

Vividly I remember my first encounter with a garfish which occurred when I was seven or eight years old. My mother and I were fishing for crappie in a lake near Fort Worth, Texas, and we were enjoying considerable success with the time-honored method of cane pole, cork, and live minnows. A longnose gar took my minnow, and as I pulled up on the pole, the hook barely took hold in the corner of his mouth. We were both very excited as the fish fought to throw the hook, and I was dumbfounded when an old man who was fishing on the other end of the dock suddenly rushed over, grabbed the pole from my hands, and hurled it well out into the lake. "Only thing to do when you catch onto a snakefish," he explained. Further explanation revealed he firmly believed a gar was a "cross" between a rattlesnake and a carp and was quite dangerous. Of course, this is impossible although the superficial coloring of the scales of a western diamondback and the scales of a gar are somewhat similar, and probably formed the foundation for his belief. To this day, I have never heard a gar referred to as a snakefish again, but it is reasonable to assume that many similar beliefs are held throughout the extensive range of the species.

For example, the roe of garfish is said to be poisonous by some fishermen in central Texas, but I have not been able to elicit an opinion on this matter from anyone in this region. To the question are garfish eggs poisonous, one Eastern Shoreman replied, "Any sumbitch crazy enough to eat gar eggs deserves to be poisoned!" Actually, there may be a grain of truth in this belief, because as far as I know, the toxicity of the eggs has not been fully investigated by fish researchers. At any rate, I see no possibility that garfish roe will replace at anytime in the near future the eggs of the sturgeon and some other fish relished by many as caviar.

In some areas of the south and southwest the so-called alligator gar abounds and may reach a length of eight feet and weigh a hundred pounds or more. These proportions have given rise to the common belief that this type of gar will attack a man and drag him under in much the same way as an alligator or crocodile. A bonafide attack on a human by a garfish has never been verified, but one can easily understand how this belief could be held. This species of gar has a blunt, short head with large alligatorlike teeth and superficially (in the water at least) resembles it namesake. It does not take an overactive imagination to conjure up a vision of one of these huge creatures behaving like a crocodile or alligator and dragging an unsuspecting swimmer beneath the depths to devour at its leisure.

Lepisosteus osseus, meaning scaly bones in Latin, is the scientific name of the longnosed gar which ranges from the Great Lakes west to South Dakota, east to Vermont, and south to Alabama, Florida, and northern Mexico. It reaches its greatest abundance in the streams, rivers, and impoundments of the Mississippi Valley. It is quite numerous in the tributaries of the Chesapeake Bay and is the only type of garfish found in our region. The long and slender head easily distinguishes it from its cousins the alligator, short-nosed, and spotted gars. The longnose gar may reach a maximum length of 120 centi-

Garfish Marauders

meters (approximately four feet) and weigh in the neighborhood of twenty-five pounds.

These fish represent an ancient lineage in the fish family tree that has not changed much in the last four hundred million years. During this vast period of time most species of fish (and other animals as well) have been greatly modified from the original forms of their ancestors. Countless other types have made an appearance on earth, lasted a few million years, and, being unable to adapt to changing environmental conditions, faded away into extinction leaving no descendents. When consideration is given to their anatomy, one can understand why the garfish have been able to withstand the rigors of drastic environmental change without significant modification.

In the first place these garfish are admirably protected by a formidable suit of armor made of so-called ganoid type scales. These primitive structures are modified into hard, polished, rhombic plates that fit tightly edge to edge to enclose the body of the fish. It is not surprising to note that the skin of garfish was used by resourceful pioneers for a variety of purposes. The gar epidermis could be fitted around the lower legs for protection against snake bites. It was also used to cover wooden plows. Indians used the scales for arrow points, and certain savages of the Caribbean are reported to have used the armor as breastplates during battles with their enemies.

I can well recall my first true realization that the gar was protected by an extremely tough covering. Another teenager and I were fishing for catfish with trot and jug lines in the Brazos River of north central Texas, and we had just completed baiting the many hooked line from our canvas covered canoe. In fact, I was just placing a large redhorse minnow on the last hook when a strong pull from somewhere up the line drove the point of the hook into my hand past the barb. The pain I suffered in cutting out the embedded hook was somewhat alleviated by the knowledge we had hooked a good size fish who was waiting on the line for us to come back and claim him. Backtracking along the line we discovered an eighteen-inch longnose gar that had managed to entangle his teeth in the drop line of one of the hooks. We were furious to find a gar instead of a cat, and, in a moment of rage, I withdrew my sheath knife and attempted to stab the beast between the eyes. The knife struck the hard scales and ricocheted back toward the canoe. A five-inch laceration of the flimsy canvas below the waterline resulted, and the Brazos flowed into the craft immediately. We beat a hasty retreat back towards the sandbar serving as our base of operations as the canoe filled with water. We barely made it to safety as we were on the verge of going under when the keel hit the gravel bottom, and we were able to jump out and pull it ashore. After emergency repairs had been made, and with my hand throbbing like a trip-hammer from the hook wound, we returned to the line. Much to our delight we found the gar had liberated himself and vanished back into the depths of the river.

The second advantageous anatomical adaptation these fish have is a swim bladder which is richly supplied with blood vessels. This organ can function as a primitive lung and acquire atmospheric oxygen for use by the animal. When the dissolved oxygen content of a body of water becomes lowered for various reasons, and the supply absorbed by the gills is insufficient to sustain normal activity, the gar rises to the surface and takes in a bit of the ozone for his oxygen starved cells. This practice can be observed readily by anglers on the prowl for these fish. Frequently several bubbles of air from the swim bladder will be given off just beneath the surface as they recharge their blood with the life-sustaining oxygen. This supplemental means of respiration gives the gar a tremendous advantage over other fish who lack the necessary equipment in an oxygen-poor ecosystem. The gars are able to survive long after most of the others have perished. The armor and the lung evolved early in the family history of garfish, and, being so beneficial to the species, were retained by countless generations from prehistoric times to the present.

Garfish are an underrated game fish that most sportsmen will pass up in search of other, supposedly, more sporting

Garfish Marauders

types. This widespread practice is largely due to the belief that gars are "trashfish". I have never fully understood what a "trashfish" is, but I think it means that a particular fish is unfit for the frying pan. Gars are also shunned because of unsupported beliefs they may be poisonous and quite dangerous. In addition to the thrill of the catch, the gar fisherman aids other species of fish each time one of the "fish eating machines" is extracted from the water. As noted previously, all gars feed only on other fish, and a large one can easily devour several pounds of food each day.

The prey is approached stealthily and quickly seized by those terribly sharp and numerous teeth. The unfortunate victim is rapidly dispatched and swallowed whole. A newly hatched garpike is born with a ravenous appetite but does not begin feeding the moment it bursts free of the egg. Rather, it attaches to a submerged root or stump with a peculiar sucker-like structure on its head. Within a day or so, however, when the young garfish has become acclimatized to its new surroundings, the sucker relaxes and the small predator starts out on its mission of death. The young gar eats so much, in fact, that at the end of the first year of its life it will be a foot or more in length and weigh several pounds. There is a record of a small two-inch specimen which killed and devoured sixteen small minnows during a twenty-four hour observation period in an aquarium.

As to the culinary quality of garfish meat, one has to remember what is nauseous to one is ambrosia to another. Gars are eaten in many parts of the country (not by many in the region of the Chesapeake Bay) and can be prepared in any conventional manner. The tough covering presents a problem to the novice garfish-skinner, but with a heavy, sharp knife and a pair of pliers, the practiced hand can denude one of these beasts with no more trouble than is experienced in skinning a catfish. The tip of a sharp knife is usually inserted into the ventral vent, and a midline cut is made as far forward as possible. After the internal organs are removed, the blade of the knife is used to detach the muscles from the skin on each side. When this procedure is accomplished, the armored plating may be peeled back toward the dorsal surface and easily removed. The meat is white and firm and is especially good when deep fried or smoked over hardwood coals.

Professor Jack Ransbottom, a biologist at Salisbury State College, recalls having dinner once with a Lithuanian family in northern Michigan. The entree consisted of fish cakes made of ground up gar meat and onions. As Ransbottom remembers, this recipe was "the specialty of the house" and made a marvelous repast.

Biologist Luther Holloway of Vicksburg, Mississippi, is the son of a commercial freshwater fisherman and is wise in the ways of preparing fish for human consumption. Dr. Holloway tells me a concoction known as "Gar Boulet" routinely graces the table of the inhabitants of the "Cajun Country" of Louisiana where he was reared, and he has kindly supplied me with the recipe for preparing this dish:

> Boil fish until tender (you may want to add garlic since the smell of the fish may be strong); add salt and pepper while boiling.
> After fish is tender, pick the meat from the bones, removing gristle. The flakes of meat should be small.
> Add about two tablespoons of flour and mix well to make the fish stick together.
> Add onions, bell pepper, celery, and any other seasoning you desire. Mix well and roll into balls. Place in skillet and panfry until done to taste.
> Make roux (flour and oil) into a thick gravy. Add gar balls and simmer for 1-2 hours (do not stir) and serve over rice.

I must admit I have not yet had the opportunity to prepare Gar Boulet à la Nanticoke, but it does sound delicious. The only thing that worries me a little is the strong smell Holloway mentions. I have never witnessed a strong smelling garfish, but, on the other hand, I have no experience with them from that part of the country.

Garfish Marauders

I think the major reason more local people don't try for garfish with rod and reel is because the hard, bony mouth of the gar makes it almost impossible to set a hook. If the most widely accepted definition of a game fish is one in which the fish will readily strike an artificial lure, a gar must be classified as game fish since it will strike a lure resembling a baitfish as readily as any other. Many times I have cast silvery spoon lures into schools of garfish and nine out of ten times would receive a rod-jolting strike for my effort. It is true, nevertheless, that for every hundred strikes, maybe only two or three fish will be hooked and subsequently landed. A game fish is also supposed to put up a battle after it is hooked. Anyone who has ever hassled with a garfish will attest to its fighting ability which endures longer than most other fish highly praised for a competitive spirit. I have the remnants of a fine glass bait casting rod in my attic that is split from end to end as a result of a confrontation with a garfish—and I must say I enjoyed splitting it very much.

Many times a gar is foul-hooked in the eye or in a soft spot on the underside. When this occurs, the wound apparently greatly reduces the fish's will to survive. With their fighting ability impaired, such foul-hooked fish are probably the source of the persistent stories one hears that gars offer little resistance to the backbone of the rod.

At this point perhaps the reader is wondering how I was able to latch onto the dozen or so I described earlier. My success in gar-angling can be attributed to the use of an easily made wire noose that is baited with a live fish. I was first introduced to the art of "gar-noosin" many years ago by an elderly commercial fisherman on the same Brazos River in Texas. When this old-timer was not running his trot lines and removing catfish for the market, he would noose gars. He had no trouble disposing of his catch to individuals or to the small grocery stores and markets in the area where the fish brought ten to fifteen cents a pound undressed. A ten-pound fish would usually fetch a dollar or more and that was not bad in the early 1940s. The natives of that wild and beautiful region of Texas were fond of gar smoked slowly for a day or so over live oak coals.

The noose apparatus is made of a pliable wire which is fashioned into a circle or noose into which the hook dangles. The garfish approaches the minnow-baited hook and spreads his fearsome jaws in anticipation of ingesting his next meal. If the angler is lucky, one of the jaws will pass into the noose (The baited noose is usually fished about two feet under a rather large float), and the other will remain outside. When the strike is made the noose is drawn tightly about the jaw and often becomes wedged in among the teeth. The fisherman must wait until the minnow is on its way down the gar's gullet (the float goes under) before striking to draw the noose tight. The gar will take a few seconds to kill its supper (the float moves abnormally along the surface) before the act of swallowing occurs. If one strikes too soon, the noose is apt to be shaken off by the enraged fish. During the ensuing action the line must be kept taut at all times lest the noose slip free. Remember, it is the circle of wire that catches the fish and not the hook. None of my acquaintances along the Nanticoke had ever heard of this gar noose before I showed them one and described its operation. Only one person exhibited any interest in the device and that interest was directed to the long-shanked hook hanging down in the noose. He wondered where I had purchased that type of hook, as he allowed a smaller version of it might be good for hardheads with their relatively small mouth.

In addition to offering sportfishing at its best and providing delicious protein for the table, garfish and many other predators are considered by some aquatic biologists to be useful in keeping the populations of prolifically spawning species under control. Without such predators as the garfish, they maintain, certain species would reproduce so rapidly there would not be sufficient food to go around. Those not dying of starvation would be stunted, and, in some cases, incapable of normal reproduction. I am not inclined to go along with the experts when they claim the predatory mode of nutrition of the garfish is beneficial to the prey populations. This may have been true two hundred years ago when the only

Garfish Marauders

menace to fish in the Bay and its tributaries was the presence of predatory fish. Today there are so many man-made conditions affecting the fish (from gill nets to pollution) that many species are having their numbers drastically depleted. I am convinced that the rockfish, trout, and others which are in trouble at this time would be better off if they didn't have a band of garfish marauders snapping at their tails.

Biologists have also found that the gills of garfish are especially attractive to the glochidia or larvae of the freshwater clam. These miniature bivalves spend the early part of their life cycle attached to the gills and gill covers of gar and other fish and are thereby distributed far and wide over the river and Bay.

Though most people turn up their noses in disgust at garfish when it is suggested they might try to catch and eat them, those that have managed to break through the barriers of fear and superstition have experienced sporting and gustatory rewards equal to any.

6

The Poor Man's Tarpon

"FISH ON!" shouted one of the dozen anglers fishing from the bridge spanning Wetipquin Creek, and he quickly added, "It surely tis old grandpa blueback this time."

I watched as the man skillfully played the fish, and I noted he held the rod tip well over his head in order to keep his small diameter spinning line taut. In half a minute the battle was over, and two identical ten-inch blueback herring had been scooped out of the water with a long-handled net. "I should aknown it was a double and not ole grandpa," the fisherman mumbled as he deftly flicked his rig, consisting of two worm-baited hooks, back into the current of the creek. As he began the retrieve of his line with an erratic pumping motion of the rod, a teenager a few feet away grunted with satisfaction as *his* rod suddenly arced into a semicircle. He too quickly landed a double and tossed them into a bushel basket that was almost full of the silvery, forktailed fish.

Everywhere, up and down the many creeks and guts that empty into the river, the same scenario was being enacted that day by countless individuals enjoying the annual spring herring run. People have been doing this every spring for as long as anyone cares to remember. It is interesting to reflect that this fast and furious action was also taking place for almost the entire length of the Atlantic seaboard at the same time in hundreds of other rivers, creeks, and streams that flow into the ocean.

The Poor Man's Tarpon

The blueback (also referred to as the summer herring, glut, or blackgut) is a member of the large herring family of fish. One way it may be distinguished from the other members of the group is by the fact its peritoneum (the membrane surrounding the internal organs of the body in all higher animals) is bluish-black in color—hence the common name of blackgut. The family includes the various species of shad, anchovies, sardines, menhaden, and alewives and probably is caught in greater poundage and numbers than any other group of fish inhabiting either fresh or salt water.

Most of the herrings taken as a result of commercial operations end up as the chief component of certain fertilizers. On the other hand, you just might encounter a pickled or smoked blackgut on an hors d'oeuvres platter at your next social function. In this part of Maryland herring taken from the Nanticoke and other Bay tributaries are put to a variety of uses. A large number are smoked or salted for human consumption, and the old practice of pickling seems to be enjoying a revival of its popularity in bygone days. Pickling has the advantage over other methods of preservation in that the acetic acid (vinegar) used serves to decalcify the many small bones, and the fish may be eaten without fear of having a bone wedge itself in the throat. Some people crisp-fry large and small herring and simply chew the softened bones to the point where they may be swallowed with impunity. My taste buds prefer herring that have been smoked with hickory or oak chips for several hours or until they have turned a golden brown.

Although herring is quite popular on the dinner table in homes along the Nanticoke, in one prepared form or another, there are many people here who will not eat them for some reason. Nevertheless, herring are put to a number of other uses. During the spring when the mighty rockfish is making its annual spawning run up the river, the inquisitive angler may examine almost every striper stomach and find it stuffed with herring of various sizes. Once I looked into the stomach of a twenty-six-inch striper I found dead on the beach near the cottage and was able to identify the remains of at least fifteen herring in various stages of digestion. Later in the spring, when bluefish and trout are caught autopsies of their stomachs will reveal these fish had an equal fondness for the so-called silver bullets. Other members of the herring family may be found inside larger fish, and it is safe to say the herrings and their relatives are of considerable ecological importance as converters of plankton to fish flesh. Because they are relished as food by game fish, they are a favorite bait of rod and reel anglers who seek the rockfish, bluefish, trout, and black perch along the river.

When the savory crustacean—the crab—appears in numbers about the first week or so in June, bluebacks are used by many residents of the river to bait their crab pots. (By Maryland law each waterfront property owner is permitted to place two wire crab pots or traps no more than 100 yards from the shoreline in front of his property, and I know of only a few residents who do not take advantage of this law which permits them to trap crabs in considerable numbers without a license.) However, during the active crab season many of these residents are scrambling around looking for suitable bait for their pots. For several years now I have had no problem with the supply of crab bait required to keep the traps in business for the entire season. Toward the end of March when the herring run starts, I place my 100 yard gill net between two poles implanted in the river bottom a hundred or so feet off shore. Within a few hours great numbers of herring have usually entangled themselves in the meshes of the net and are awaiting removal. It is not infrequent that I have caught 50 to 150 herring during a twelve-hour period. Most of the catch is placed in two large plastic barrels which contain a supersaturated salt solution and is stored in the basement. A few of the fish find their way to the pickling bath or the smoking rack, but most go into the salt where they will remain preserved for an indefinite period of time.

When my crab pots are set out about the first of June each year, they are baited with the preserved herring, and I usually

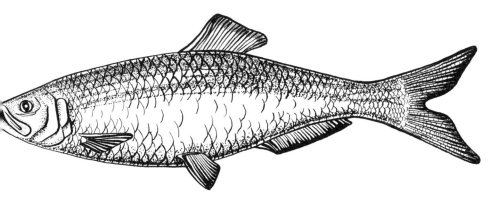

Blueback herring

have a sufficient supply salted away to last until October when traps are removed from the water for winter storage. Some crabbers claim other types of bait, such as chicken necks and backs or fresh fish, will lure more crabs to the pots than salted herring, and they are probably right. But, my preserved herring bait is always available at practically no cost (the salt bought in twenty-five pound bags is the only expense incurred). And, I always seem to be able to entice enough crabs to my pots with herring so that by the end of the season I much prefer an old-fashioned hot dog to a lump of backfin meat. During the summer I frequently donate excess crabs to friends in Salisbury.

Herring are utilized to generate food for human consumption in another way. Many along the Nanticoke coordinate the planting of their spring gardens with the annual herring run. Two side-by-side furrows are made for each row of the garden. The seeds are planted in one furrow and herring in the other for fertilizer. Thus, the fish indirectly contribute a great deal to the gardener's nutrition as the vegetables he harvests each season are impregnated with the minerals and other vital elements that once made up the protoplasm of the herring. Actually, the use of herring and other fish to fertilize crops is not unique to the Nanticoke River region but is a common practice throughout the Chesapeake Bay region. The Indians probably related this method for improving crop yield to the early settlers.

The runs occur each year because, like the rockfish, the herrings are an anadromous species, which means they leave the ocean and swim upstream into fresh water during the late winter and early spring to attend to reproductive functions. The amazing thing about the annual herring migrations is that each fish returns from the sea to reproduce in the same stream or creek where it was spawned.

In late winter millions of herring far out in the cold depths of the sea appear to respond to certain events—events such as ocean currents shifting a bit, the temperature of the water warming up a degree or so, the period of daylight lengthen-

The Poor Man's Tarpon

ing, the earth tilting a fraction on its axis—and seem somehow to know it is time to head for shore. The youngest of these migrants has been at sea for four years while maturing sexually, and now their gonads are releasing the chemical hormonal substances that direct the reproductive instinct. Older fish will receive the chemical message as well as the physical stimuli and will start the long trek for the second, third, and for a few, the fourth time.

Authorities believe there are as many physiological varieties or tribes of herrings as there are spawning areas. The fish are known to respond to such physical phenomena as latitude and longitude, polarized light, and the magnetic field of the earth. These factors, coupled with their ability to detect minute changes in salinity, temperature, and current oscillations, guide them to the mouths of the ocean tributaries. In the meantime the chemicals from the various reproductive areas have arrived at the mouth of the estuary. From here on it is thought the unique biochemical quality of their birthplace takes control of those who react positively to the stimuli. A fish not engendered in a specific environment (which may be only a ten-foot wide rivulet or gut) will not respond. An indelible path to the breeding site is set forth, and the instincts within the fish are so strong it cannot deviate from its determined route.

The complexity of the inheritance of these instincts (as well as the natural selection processes that implanted them into the genetic makeup of a particular species in the first place) is enough to overwhelm the mind if too much philosophical thought is given to the matter. But, it does seem obvious that additional research on these processes could provide information leading to a better understanding of all animal behavior—perhaps even man's.

It is easy to understand how this delicate response system may be upset or obliterated when nature, in the form of a hurricane, dumps untold millions of gallons of water on a reproductive site at a critical time in the spawning season. The chemical, and, to some extent, the physical signals or markers of the nursery may be so diluted they cannot be detected by the herring. When this occurs, there is a poor or nonexistent spawning run. This happened when Hurricane Agnes struck the Eastern Shore a few years ago. Trillions of gallons of fresh water were deposited in the Chesapeake region. In some areas, oysters and crabs were almost wiped out due to a drastic decrease in salinity, and the natives along the Nanticoke remember there was hardly any herring run at all that year.

Recent studies with anadromous species (including the prized rockfish and shad as well as the herring) have shown chemical pollutants can greatly upset the spawning spoor. It seems to me chemical pollution by pesticides, fertilizers, and industrial wastes of various types is probably more important in upsetting the biochemical pathways than are natural phenomena. In some years herring and shad runs did not occur, and there was no natural episode on which the blame could be laid. However, to my knowledge, specific agricultural or industrial pollutants have never been pinned down as the cause of the disruption of a specific spawning run; although this very well may have happened, it is extremely difficult to prove.

When all the natural forces are meshed together, the instincts of the fish read GO, and the annual run begins. It does not take long for the word to spread among the blueback enthusiasts who have been waiting patiently all winter for some piscatorial action. Usually the first sign of the run is a dramatic increase in the number of herring taken in gill nets such as used by the property owners or the commercial fishermen. As the fish move into the creeks and streams, they are so numerous the water is roiled with their movements and gyrations. In the smaller rivulets the herring runs have been compared to a huge mass of quicksilver pulsating through the water. At the upper reaches of Wetipquin and Quantico creeks the herring are dipped from the water in tremendous numbers with a net attached to a long pole. I have seen the water in the branches of the Wetipquin literally turn white because of the presence of the multitudes of silver bullets which some refer to as the "Poor Man's Tarpon." Bushel basket after bushel basket is filled from the dip nets and is

The Poor Man's Tarpon

carted away by the natives to be salted for crab bait or used to enhance the productivity of the earth.

A generation ago few anglers sought the herring with rod and reel. Dip nets of every conceivable type and shape were used almost exclusively, and, as indicated, are still employed today by those who want large quantities of fish. With the advent of spinning and spin-casting fishing equipment not too many years ago, it did not take long before the sportsman discovered herring were ideally suited to this type of light tackle. They were easy to catch, occurred in large numbers, struck artificial lures readily, and fought like a tarpon, albeit, if only for a short span of time. Herring, unlike salmon and some other anadromous species that will not take the hook during the spawning run, seem to have no qualms about stuffing themselves even while in the actual process of releasing eggs and milt.

Almost any type of natural bait will be taken. A medium size redworm may be cut up into three or four separate baits, and a small squid cut into narrow strips can provide enough bait to last almost an afternoon. Doughbait or just plain bread kneaded onto a number eight hook is popular with youngsters and is quite effective. Another tidbit prized by the blueback is ocean shrimp which may be purchased frozen or fresh in the fish market.

A light action rod with a line testing no more than six pounds offers the most sport with herring which seldom weigh more than a pound. The usual technique is to affix just enough split shot a foot or so above the hook to enable a cast to be made with ease. The line is then allowed to sink to varying depths and is retrieved in a slow erratic motion. It is best to start at the top and work the bait at different levels until the zone of most activity is located. On rare occasions the best results may be obtained by a rapid and steady retrieve, but most frequently the slow, jerky method will prove to be the best.

A majority of fisherpeople have discovered artificial lures will do just as good a job (if not better) than any of the natural offerings. I remember the first time I fished for herring from

Wetipquin bridge. I was the only one of some ten anglers who was casting artificial lures. My lures caught many more fish than any of the assorted worms, grubs, and cut bait used by my neighbors. Many were amazed at this, and gathered around to observe my technique and to ask questions.

The most popular artificial lure for herring or shad is the red and white shad dart sporting a short white bucktail. These inexpensive and effective lures cast easily on light tackle. A common terminal rig will consist of a couple of split shot on the line for casting facility and two darts a few feet below. One of the lures is tied to the end of the line, and the other is attached to an eighteen-inch dropper line about two feet away. The contraption is cast into the current and retrieved (as in bait fishing) in a variety of ways. A sort of jigging and pumping up and down motion is frequently employed with the line being rapidly reeled in each time the tip of the rod is lowered. When the strike (or strikes) comes, the hook must be set immediately with a firm but not jerky motion. The mouth of the herring is soft, and one can expect to lose one out of every three fish he hooks properly. As noted, a double strike is common, and, with one fish dashing off in one direction and the other going another way, a person has his hands full preventing a momentary line slack and subsequent thrown hook. Of course, one way to overcome this problem is to use only one lure.

Other artificial lures will often be found to be as effective as the shad dart. Small, inexpensive, silver-colored spoons are excellent as well as small spinner baits such as the Colorado and June bug varieties. I have good luck with homemade lures constructed as small as possible and which resemble the Abu or Shyster commercial models. These lures cast well, and, in addition, I have the satisfaction of catching fish on lures I made myself. This is good for my ego—to outsmart a fish on a homemade lure. All too often I get the distinct impression the fish are smarter than I am.

For me, the pinnacle of achievement in herring fishing is the taking of these bits of quicksilver with the fly rod. A shooting line is usually necessary to get the fly to the area

The Poor Man's Tarpon

where the action is taking place. Streamer flies (which resemble small fish) are most frequently used, but very small spinner baits are very effective if the angler has the necessary skill with a fly rod to handle them. Though the total number of fish taken with the fly rod may be fewer than if other types of equipment had been used, the long and limber switch truly makes a lowly blueback into a poor man's tarpon.

7

Cats, Toads, and Stingrays

OF ALL THE FISH inhabiting the Chesapeake Bay the catfish, toadfish, and stingray are the only ones possessing a venom-injecting apparatus capable of causing a painful and dangerous wound in man. As with reptiles, mammals, and other forms of animal life, many species of fish are unjustly accused of being dangerous to man. Actually only a few fish throughout the world pose any threat to the human homeostasis. There are no fish of this ecosystem which have poisonous flesh unsafe for human consumption. Numerous stories are repeated about odd looking fish such as the sea robin or toadfish which should never be eaten, reportedly because the meat has a potent toxin incorporated in it. However, when one of these story tellers is pinned down he will usually admit that he has never known directly of anyone being harmed by the meat of these fish, but that a friend of his told him of a case several years ago. The sea robin, incidentally, is only found at the mouth of the Bay and in the open ocean. In recent years there have been reports in the local newspapers dispelling the myth that the sea robin is poisonous and, in fact, have stressed it as a gourmet's delight. I showed these accounts to a friend who had been adamant in his insistence that the sea robin was poisonous, and he backed up a bit, stating he had meant that the eggs of the robin were very toxic. As with the eggs of the garfish, there may be some truth in this belief, but as far as I know, no scientific investigation of the matter has been undertaken.

Cats, Toads, and Stingrays

Catfish

On the other hand, I have never heard anyone suggest that the meat of catfish was dangerous to the human physiology. In fact, the majority of the people in the United States who are regular fish eaters consider catfish to be one of the finest tasting types of seafood you can place on your table. In verification of this, you only have to travel through any of the southern states and observe the signs posted by restaurants and diners along the road proudly proclaiming the specialty of the house as "Fried Catfish" or "Catfish and Hush Puppies." Many of the more elegant dining establishments do not advertise their fare so boldly (I once saw a flashing neon sign by a diner near Kentucky Lake which advertised its cats were the best in the region and the glass representation of a catfish alternately winked its eye at you and wiggled its whiskers), but somewhere on the fancy menu you are sure to find a catfish dinner listed with all the trimmings.

Providing catfish for the market has become a major source of income for many residents in the southern states. A generation ago the supply of catfish was amply provided by commercial river and lake fishermen, but, for a variety of reasons, the tonnage of cats harvested from our natural waters has dwindled and is not sufficient to meet the demand. To cope with the shortage many farm owners have converted a few of their acres formerly devoted to peanuts or cotton production to rearing catfish. In most cases this change proves financially rewarding as pan-sized catfish, dressed and ready for the skillet, bring as much in the supermarket as a good grade of beefsteak does.

Every outdoorsman has tucked away in the recesses of his mind, but always available for instant recall, memories of certain past experiences he has enjoyed very much. If others are like myself, the slightest encounter with a stimulus that could be associated with these experiences is enough to bring them back into sharp focus—if only for a minute or two. Whenever I see anything associated with catfish, my mind instantly flashes backwards about thirty-five years to picture vividly several pleasant experiences. The most exhilarating of these episodes occurred when I was a Boy Scout under a scoutmaster fortunately both devoted to his boys and to going fishing. This scoutmaster, also the proprietor of a small grocery store in my native north Texas, frequently became afflicted with fishing fever. The only thing to alleviate the condition was to "get ahold of some of the boys" and head for the Brazos River for a night of camping and catfish trotlining. His customers would note the "Gone Fishin" sign on the door and usually wait for his return to make their required purchases.

Normally we would arrive on the river in the late afternoon and immediately proceed to seine several buckets of minnows to be used as bait. Just as the skies were darkening, two or three trotlines with at least fifty hooks dangling from each had been set out and baited. We would then set up an informal campsite on the bank of the river and light an enormous bonfire. Every hour during the night the lines were checked, and the numerous catfish found to be hooked would be removed. Somewhere around two or three in the morning our scoutmaster would build up the fire and produce the largest frying pan I have ever seen. (The pan was at least two feet long and a foot wide and was constructed from an old steel boiler he had found discarded years before.) As he was melting a couple of pounds of lard in this oversize skillet, the boys would know their job was to select a dozen or so cats in the one pound class and properly skin and prepare them for cooking. Our leader always cautioned us to be careful not to let one of them get a fin into us, but sometimes this could not be avoided. He insisted the proper way to kill a cat was to make a small hole with your knife between the eyes and then insert a straw or twig into the brain. I never understood why this method was more effective than simply chopping off the head, but we always dispatched the fish in this manner.

When the grease was sizzling, the fish were patted in cornmeal and dropped into the pan which we referred to as "Big Bertha." Previously prepared hush puppies made of cornmeal

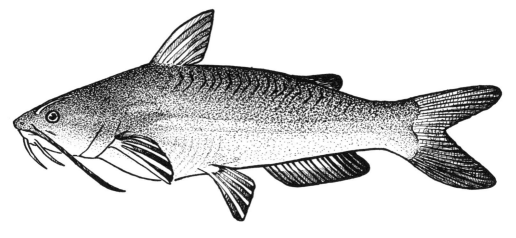

Catfish

were placed in the smoking fat at the same time. In a minute or so the cooking was completed and the fish and "pups" were allowed to drain a bit on a wire grate. I have dined in four-star restaurants in Europe and in the major cities of this country on a variety of award-winning meals prepared by famous chefs. However, I cannot recall any of these repasts as more enjoyable than the catfish and hush puppies prepared in the big pan on the banks of the Brazos River in the middle of the night by our scoutmaster. Undoubtedly, the star-studded Texas sky, the comradeship, and the overall good feeling added to the sensations as the meal was consumed, but, even today, when reminded of those outings the taste experience comes back as strong and delightful as it was thirty-five years ago.

In spite of the fact that everyone along the banks of the Nanticoke knows the meat of catfish is not dangerous to man, fully seventy-five percent of the inhabitants will not eat it for love nor money. Lou Griffin, a river resident since 1950, tells the story of the owner of a plush restaurant in nearby Salisbury who used to make weekly trips to the Nanticoke to catch large numbers of catfish. After proper disguise had been added by the cook, these fish were placed on the menu as "Chesapeake Perch" and were a great favorite of the epicures of Salisbury and surrounding areas for many years. Someone told on this enterprising restaurateur one day, and the requests for Chesapeake Perch rapidly dwindled to the point where it had to be eliminated from the menu.

There are twenty-four different types of catfish found in the waters of the United States. Every state, with the exception of Alaska, has at least a few of the different species represented. All species of catfish look alike, and in some instances it takes a trained eye to distinguish one type from another. Catfish have eight whiskers, or barbels as they are called technically—a large one trailing from each corner of the mouth; a shorter pair beneath the nostrils; and four under the chin. They have large flattened heads and wide mouths containing many small, inoffensive teeth. The skin is scaleless, but they do have three large, sharp pointed spines associated with their anterior fins. Each of the lateral or pectoral fins has one of

Cats, Toads, and Stingrays

these spines on its leading edge as does the single dorsal fin on the top of the body.

As noted, people tend to distrust *any* animal having an uncommon or abnormal appearance, and this is certainly true of the rather ugly picture the catfish presents. But, more than his outward appearance, most individuals tend to shy away from him because they know those spines incorporated into the fins spell trouble.

Located at the base of the pectoral spines are loose aggregates of venom-producing cells that secrete a toxin to the outside via a small pore. As the spines are held parallel to the body, they are bathed in the material as it oozes from the body. Contrary to popular belief, the spine is not hollow like a hypodermic needle as is the case with the fangs of pit vipers. It is only the pectoral fins that have the venomous apparatus; the dorsal spine lacks such an arrangement in all catfish in the United States except the small, so-called madtom which is an inhabitant of the Midwest. When the shoulder or pectoral spine is introduced into the skin some of the fiery venom is carried into the wound as well as a good supply of bacteria-laden mucus from the skin. Fortunately for us, the poison has been diluted greatly by the water as it enters our bodies. Some venom studies have shown that the venom of a catfish—in an undiluted state—may be as potent as that of a copperhead snake. It seems to me the exact nature and action of catfish venom would be an interesting and perhaps beneficial research project for some venomologist. Venom from cobras and vipers have been found to alleviate several human ailments when processed into therapeutic preparations.

The catfish we have along the Nanticoke is the white catfish which is called the Potomac cat or catfish of the Potomac by residents of the Western Shore. The white cat is very abundant throughout the Chesapeake and is, in my opinion, the most numerous food fish we encounter at all seasons of the year. The white cat is sometimes confused with the similar channel catfish which is found in the upper reaches of the Potomac River and the extreme northern part of the Bay itself. Rarely do they venture as far as the mouth of the Nanticoke or

Tangier Sound. Both have a forked tail, but the channel cat has an irregular series of dark spots on each side which the white cat lacks. There are about twenty-two rays in the anal fin of the white catfish, whereas the channel catfish has twenty-four or more. In addition, channel cats routinely grow to a weight of several pounds while the average size of the white catfish is about one pound. White cats in the range of two to four pounds are fairly common, however, but anything weighing more than this is quite rare. The largest white cat ever taken tipped the scales in the neighborhood of thirteen pounds.

Years ago the white catfish was found only from the Delaware River along the East Coast to Florida to the Gulf of Mexico, but now it has been introduced into all of New England and California as well. It has always been recorded as a common inhabitant of the Chesapeake Bay and associated feeders.

To those of us who have no aversion to eating the white catfish it represents an excellent source of protein which is easily procured and quite tasty. It is readily taken on rod and reel and is a frequent visitor to gill nets. Almost any sort of bait may be used including live bait fish. One writer observed the stomach contents of several white cats he had taken and found two to three inch crappie in most of them. Many times in recent years I have absolutely refused to pay the outrageous price demanded for blood worms to be used as bait for the glamorous rockfish of the Nanticoke. Bloodworms are always the best bait for rock, and at certain times of the year they apparently can be caught on nothing else. Refusing to go along with the bloodworm ripoff (in 1979 a dozen scrawny bloodworms were selling for about three dollars!) I have passed many an enjoyable day angling for catfish using nightcrawlers dug from my backyard. With these as bait, it is rather satisfying to return from a trip on the river with a "boatload" of catfish to show for my efforts and observe the sophisticated (but frustrated) rockfisherman cussing wind and tide, the Natural Resources Department, and a variety of underwater gods, because the prize rock has not taken his expensive bloodworms.

Cats, Toads, and Stingrays

Cats should be skinned before preparing them for the table. This practice is not absolutely necessary, but I have found they invariably taste better if the mucus-laden skin is stripped off with a pair of pliers before cooking. Cats may be prepared for the table in countless ways, but my favorite method involves the smoking of fillets over a low charcoal fire to which a few chips of hardwood have been added. I amply salt and pepper the pieces of fish and brush on cooking oil during the smoking procedure to insure they will not become dry and tasteless. Once I offered a piece of this firm, sweet tasting, white meat to a neighbor who abhors catfish. Upon tasting it he replied, "Best damn piece of rockfish I ever put into my mouth."

There is a drawback, of course, to catfishing. Those pectoral spines must be respected at all times. I know of no one in our region who has not been "finned" at one time or another in his life, and they all relate a similar story as to the consequences. The initial sensation is one of intense burning and pain that may last for a considerable time. The area around the wound pales immediately after the sting, but the pallor is soon replaced by a cyanotic appearance. Redness and swelling soon follow. The wound made by the spine is ragged as the spear has a series of rough recurved teeth on its backside. Considerable damage is done when the spine is pulled free of the body. Punctured and lacerated wounds like this are prone to secondary infection due in large part to the harmful bacteria that are introduced with the mucus and slime on the barb. Many catfish wounds require days or weeks to heal completely, and some authorities believe, not without supporting evidence, that there is a necrotic component of the venom that destroys many of the cells in the area of the wound.

Treatment of catfish stings is as varied as the remedies for the contacts with sea nettles. The most commonly used agent of treatment along the Nanticoke is old-fashioned household ammonia, but one does run into a kerosene advocate now and then. As a matter of fact, my old scoutmaster was a member of the kerosene group. Medical authorities have nothing specific for any type of fish sting, but you can find general recommendations for the treatment of any venomous fish wound. Most recommend the mouth or some other type of aspirator be used to suck as much material out of the wound as possible. No physician I know goes so far as to suggest the cutting before sucking procedure used in snakebite. Most doctors do recommend soaking the member in hot water for thirty minutes to an hour with the temperature of the water as hot as the patient can tolerate. If the wound is on the face or body, hot compresses are recommended. The addition of magnesium sulfate (epsom salts) to the water is believed to be beneficial, and, if possible, the infiltration of the wound area with a procaine injection may be accomplished to relieve prolonged pain. If pain persists for an extended time, some success has been obtained with the intramuscular administration of Demerol. You will be pleased to learn that the vast majority of catfish stings do not require the above mentioned procedures as they are usually no more serious than the sting of a honeybee or yellow jacket wasp.

Great care must be used, nevertheless, when removing a cat from a hook or net. When running my gill net I always have a small pair of sharp wire cutters dangling from a string around my neck. When working a cat free of the meshes I always use the nippers to snip off the pectoral and dorsal spines at the skin line before any attempt is made to get the fish free of the net. When taking a catfish off the hook (more often than not it is embedded deep in his gullet), it is best to grab the fish by the lower jaw with a pair of pliers before attempting to work the hook loose. When tormented, the cat will lock the dangerous spines in a forward position, and this little trick makes them even more of a defensive weapon. Most experienced catfishermen (unwise as it may be) have developed a procedure of handling their catch without the use of pliers. With a little practice the right hand can be carefully placed behind the spines and, with a forward pressure, they become locked in a right angle position. The grip must be firm (a cloth glove is helpful in this respect) as the skin is slippery, but grasped in this position the cat is more or less helpless, and the other hand can be used to disengage the hook.

Cats, Toads, and Stingrays

It should be noted that the venom maintains its potency for at least several hours after the fish has been removed from the water. Once my small son hooked a cat and swung him into the boat. Just as the fish struck the bottom of the craft the hook bounced loose and the fish made its way to concealment under one of the seats. Several hours later I changed positions in the boat and my bare foot stepped on the dead fish who had previously slithered from the protection of the seat. Immediately I experienced the familiar fiery, throbbing sensation that continued for the better part of an hour. That fish had been dead for an extended period as its body was stiff with rigor mortis when I stepped on it.

When dealing with catfish one has to take the good with the bad, and though most of the inhabitants of the Eastern Shore want nothing to do with them, I, and a few others, calculate the good far outweighs the bad and will continue to seek them out for food and sport at every opportunity.

Toadfish

While the catfish is not apt to win any beauty contests the toadfish (also called the oystertoad) is easily, as indicated by one author, the "Champion of Ugliness." The toadfish bears only a superficial resemblance to its amphibian namesake and is a rather small species. It rarely if ever grows longer than eighteen inches, and one weighing in at over a pound is quite rare. The toadfish has a large, flattened, grotesque head that occupies fully one third of its total body length and a mouth that stretches from gill cover to cover. Its color varies from dull greenish to brownish with dark, wormlike markings along the side. The skin is scaleless, but the surface is pitted with numerous mucus producing glands, which make the fish as hard to handle as a cat.

Because there are two sharp spines found in conjunction with the dorsal fin and one additional sticker on the rear edge of each of the gill covers, the toadfish is dangerous. In contrast to the spines of the cat, these spines are constructed like a pit viper's fang and inject the venom into the victim directly with no chance of dilution by the river water. Most residents of the Bay area fear the sting of a toadfish much more than that of a catfish. There is no indication the venom of the toad is any more potent than that of a cat, but the sting is probably more severe because of the quality and quantity injected—little chance for it to be watered down. I have never been stung by a toadfish, but the individuals I have questioned who have had the experience state the sensations are approximately the same as would be felt after an encounter with a catfish.

Three species of toadfish inhabit the waters of this country, but only one of these is found in the region of the Nanticoke and its environs. Toadfish in other parts of the world are apparently more dangerous than the one we deal with. Toads are frequently caught over the numerous oyster beds which dot the bottom of the river, and for this reason it is believed that they feed mainly on oysters. They have powerful jaws that are capable of exerting tremendous downward pressure. Coupling this fact with their fondness for oyster bars, the fish are believed by Nanticokians to crack the oyster shells and dine on the succulent contents. Toadfish have considerable strength in their jaws but not nearly enough to fracture the tough shell of an oyster. It is known these fish prefer to deposit their eggs in a shell or other nesting place, and this is probably the reason they are found commonly in the company of the bivalves. In addition, oyster shells attract a variety of marine worms and snails, and these are relished as food by the toads.

As with the cats, the toadfish usually is hooked deep down in his esophagus. Veteran anglers in this region include in their tackle box an implement made out of coathanger wire to aid in the dehooking process. The wire is "J" shaped at one end and is slipped down the line until contact with the hook shank is made. A flip of the wrist is all that is normally required to wrench the barb loose. The toad is then usually impaled on the blade of a knife and flung far back out into the river. If no such device is available the angler must proceed with caution in extracting the hook. Never permit your finger or thumb to drift inside those massive jaws. I did once and lived to regret my carelessness. The toad clamped down on my thumb and

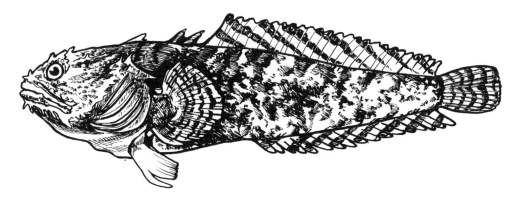

Toadfish

held it in a viselike grip. The numerous teeth are short and stubby but are not especially sharp. They do serve to anchor any object that may find its way into the mouth. Try as I did I could not force the jaws open, and I finally had to decapitate the toad. Even then, the jaws were locked in a reflex action and could not be pried apart. My only recourse was just to yank my thumb free. It came out a bleeding mess that looked to be in worse condition than it actually was. In any case, it was an unpleasant experience I have been fortunate enough to avoid ever since.

As noted previously many of the natives think the meat of a toadfish very venomous and to be avoided at all costs. To others it matters little whether or not the toad is poisonous. Its repulsive appearance deters any thought of consumption. Bob Mitchell, of Mount Vernon on the neighboring Wicomico River, is a graduate in biology from Salisbury State College and knows full well the meat of a toad is not poisonous. I asked him if he had ever partaken of toadfish fillet and his answer summed up the feelings of many: "If I was on the verge of starvation and an oystertoad was the only thing that could save me, I would rather die than put a piece of one of those ugly bastards in my mouth."

Ulysses Mentasti of Salisbury came to this country from his native Italy in 1950 and settled into a residence at Public Landing near the ocean in Worcester County. Mentasti knew of no fish in the old country that was poisonous to eat, so he relished the many toads he caught near his home. His wife made a delicious fish chowder out of them, and he especially enjoyed the broiled fillets of larger ones. It was only after two years of relishing toadfish that a neighbor told him the ugly beasts were deadly poisonous. Mentasti wondered at the fact that he was still very much alive and healthy, and to this day keeps every toad he catches for the stew pot, frying pan, or freezer. I prepared and devoured a toadfish several years ago and found it to be quite tasty. The meat was better fried than that of the much sought after white perch. Needless to add, I suffered no physiological disturbances whatsoever.

The croaker, or hardhead as it is frequently called, is a

Cats, Toads, and Stingrays

common food and game fish of the Nanticoke these days, and it well deserves its name. The croaker will produce a weird guttural sound when it is pulled from the water. However, the ability of the croaker as a vocalist is overshadowed by the sound-producing talents of the toadfish. Underwater, the toad utters two major sorts of sound—powerful foghornlike "boops" combined with shorter grunts associated with deep growling sounds. As with many other fish capable of sound production, the noise is made by the swim bladder or hydrostatic organ. The swim bladder of the toad is heart shaped and lined along each side with red muscle fibers. By contracting and relaxing the muscles, the bladder is made to change shape and the air forced out and in produces the sound effects. Toadfish sounds are so loud a man can hear them out of the water when they are made by a fish under the surface. One researcher decided to measure the strength of the sounds under water and reported he was almost deafened when he swam close to the toad. At a distance of two feet he calculated the strength of the sound waves to be 100 decibels—roughly equivalent to the force of the sounds made by a jackhammer or a subway train passing a station at full speed.

Toadfish are territorial creatures who establish themselves in a specific portion of the river bottom. Any intrusion on a toad's home range is aggressively countered. I suspect the potent pectoral spines, viselike jaws, and the ability to produce sound waves strong enough to rupture an intruder's eyeballs act as an effective deterrent to any other species which would attempt to move into the area.

Stingrays

There can be little doubt the most dangerous creature to inhabit the lower region of the Bay and its estuaries is the stingray. Stingrays are in the same class of primitive fish as the sharks and skates and are mainly characterized by the lack of any true bone in their bodies. The skeleton is made entirely of cartilage or gristle. The mouth of these creatures is ventral in position, and the openings to the gills are not covered. The mouth contains many sharp, replaceable teeth. A stingray can be thought of as sort of a flattened out shark, but it has a lethal defensive weapon the shark does not possess—a long-pointed poisonous spine on the dorsal portion of its drawn-out tail. Only the butterfly rays or sand skates, as they are sometimes called, lack a poisonous spine. This spine is bonelike and firmly anchored in the muscles of the tail. The tip is very sharp and the sides are lined with many serrated edges or teeth that are recurved back toward the base. Two parallel grooves run from the base of the spine to the tip and contain a spongy-looking cell mass which is the primary venom gland. Some venom is apparently secreted from specialized cells at the base of the spine. The terrible spine is enveloped in a sheath of skin when not in use and is consequently bathed in the venom and a film of mucus. It is not hollow like the venomous spines of the toadfish.

Most stingray stings occur when the fish is stepped on in shallow water where it has partially burrowed into the mud. The tail swings forward toward the enemy, and the powerful muscle drives the spine into the tissue. A lacerated or puncture type wound is produced. Pain is the predominant symptom of stingray sting and usually develops immediately or within ten minutes of the attack. The pain is acute and throbbing, and the wounded area becomes cyanotic followed by reddening and swelling of the area. Sometimes generalized symptoms such as fall in blood pressure, vomiting, diarrhea, sweating, rapid heart beat, muscular paralysis, and even death can result. There is no specific antidote for the poison, and the wound is usually treated in the general way described earlier. Penetration of the skin and underlying area is usually accomplished without serious damage to the tissues. The damage occurs when the stunned victim acts reflexively and attempts to withdraw the spine. The recurved teeth along the sides of the spine then tear the tissue apart as they are forcibly withdrawn. An ironclad rule that everyone along the river knows, is never attempt to pull a stingray spine loose from your body. Surgical removal of the dart is required if the damage is to be kept to a minimum. There are about fifteen hundred stingray attacks

Cats, Toads, and Stingrays

reported in the United States every year, and every now and then you hear of one taking place about the Bay. I have been unable to talk to anyone who has had this traumatic experience, but plenty of people say they know of someone who has. However, it always seems that the unfortunate person is now dead or presently residing in some distant place like Montana.

There are several types of stingrays that venture into the Chesapeake Bay at various times of the year, but in general the cownose ray is the one most apt to be encountered along the Nanticoke, Wicomico, and Pocomoke rivers and Tangier Sound. The protruding head of the cownose with a well defined indentation at its apex is distinctly set off from the rest of the body, and this feature serves to distinguish it from all the others except the eagle ray. The eagle ray has a head that is more or less straight edged, but in any case it rarely enters these waters.

The largest cownoses may reach forty-five inches across their flattened bodies and weigh more than fifty pounds. They are found in the Nanticoke from May to October and migrate to the coast of North Carolina when colder weather arrives. Often the tips of their "wings" break the surface of the water as they frolic in the shallows, and this may set off an alarm of sharks in the area. Sharks are reported occasionally, but to my knowledge they have never bitten anyone or, in other ways, caused problems.

The cownoses frequently travel in schools from ten to one hundred and may be observed feeding in certain areas where food is abundant. They prefer hard and soft shelled bivalves, mainly oysters and clams. There is no doubt that the cownose uses its powerful jaws and cementlike teeth to crush the shells of these mollusks to get at the inside. All reports indicate stingrays have increased in our area over the past few years, and this has caused considerable concern among the watermen who tong for oysters and clams. Many have voiced the opinion that the poor catches of shellfish in recent years are directly related to the population explosion of the rays.

Cownoses are frequently caught on hook and line, and if you are lucky (unlucky?) enough to tie into a big one, you can be assured of a long and tough battle. I hooked one off the Great Shoals light (over an oyster bar) at the entrance to the Nanticoke some years ago and fought the monster for well over an hour. My companion urged me to cut him free, but I had an idea of dragging him aboard the boat to take a series of photographs. When I was finally able to bring him alongside, I stood poised with the gaff in hand to haul him aboard. Just then one of those huge eyes looked squarely into mine and seemed to say, "I wouldn't do that if I were you." I glanced at the menacing spine on his long tail, drew my knife, and cut the line.

There is absolutely no reason in the world why rays and sharks are not eaten by more people in this country. In Europe they are considered to be somewhat of a delicacy, and much of the fish in the traditional British fare of "Fish and Chips" is actually shark or ray meat. Shark and ray meat are marketed in England under the name of flake. In spite of this English connection, I know of no resident of the Nanticoke who will admit to having eaten shark or ray meat. The University of Delaware Institute of Marine Science has recently published a comprehensive booklet on how to prepare ray and shark for the table and does everything possible to encourage people to utilize these species for food. But, I doubt if this scholarly publication will change the minds of many of my friends.

In some areas of the Bay the watermen use the cut up wings of rays as bait in their crab pots, and some of these commercial fishermen report it to be more attractive to the blue crab than the traditional lures of salted eel or menhaden. Of course, these watermen will not eat rays themselves. Not long ago many Americans all over the country ate ray meat without knowing what they were eating. A common practice among a few notorious seafood packing houses along the East Coast was to create "Ocean Scallops" out of stingray. A boring tool approximately the size of a silver dollar was used to gouge rounded chunks of meat from the ray wings. These long pieces were chopped into scallop-size morsels and sent on their way to the supermarket. Actually, a real scallop is the large cylindrical muscle of the common mollusk known as *Pectin*.

Cats, Toads, and Stingrays

The marketing of ray meat under the name of sea, or ocean, scallops continued until some disgruntled employee blew the whistle on the practice, and the federal government ordered the packers to stop the deception. There seemed to be little market for the fish when the package was properly labeled, and rays are no longer taken for processing.

The cownose ray is a dangerous animal that humans should approach with the utmost of caution. Rays will not bite like sharks, but the damage their stinger can inflict is sure to be extremely painful, if not fatal.

Many of the Nanticoke River people would choose to forget the catfish, toadfish, and cownose ray exist, but this is impossible as they are well established as an important part of the fauna. Though they can be dangerous, these creatures may be used effectively for the benefit of mankind if their biological backgrounds and modes of life are understood.

8

Where Have All the Rockfish Gone?

I WAS DEEPLY SADDENED when I found the majestic *Morone saxatilis* (commonly referred to as rockfish, rock, striped bass, or striper) lying dead on the beach about fifty yards from the house. She had been a real beauty in life—measuring some thirty-six inches in length and weighing in at what I guessed to have been twenty-five pounds. As she lay decaying on the sand, the hundreds of thousands of eggs oozing from the vent on her underside attested to her fertility that spring spawning season. There was no way to determine how many times she had made the trip up the Nanticoke to fulfill her biological obligation of reproduction, but, obviously, this journey was her last one.

I approached the family of bank fishermen who had my permission to park in the backyard, make their way up the beach to set up camp for the day, and implant their long surf-type fishing rods in the sand. Four generations were represented in this piscatorial clan with the patriarch, old Jessie Allen, who is somewhere between eighty and ninety years old. I asked old Jessie, wearing hip boots and standing knee-deep in the frigid water, if he had any idea of what had happened to such a prize fish. For a moment he said nothing, but finally shifted his chaw of Red Man to the other cheek and grunted as he pointed to two long stakes protruding from the water about two hundred yards offshore. One could see the stakes supported a gill net which was just visible in the ebbing

Where Have All the Rockfish Gone?

tide. "Goddamned netters killed that fish!" he retorted at last, and added, "Too big to keep so they had to chunk it back; damn net probably did it in first, anyway."

There seemed to be no question the wasted fish had been trapped by the gill net and then thrown back into the water because it exceeded the maximum length of thirty-two inches as dictated by Maryland law. To be apprehended by the Marine Police with an oversize rockfish in your possession can mean a fine of approximately five hundred dollars and the possible confiscation of your boat and fishing equipment. The fine is not so important to the waterman, as he can make that back and more with a good day's catch, but the loss of his boat and gear for an indefinite period is a strong deterrent against the temptation of retaining an oversize fish. With the waterman the economics of life are simple: no catch, no pay.

The idea of returning these superspawners caught in the nets in order to help preserve the numbers of this endangered fish is a noble one, but it does not take into consideration the fact that at least seventy-five percent of the fish taken in a gill net are fatally injured. The meshes of a gill net are designed to admit the head region of a fish that runs into them but are too small to permit the passage of the entire body of the victim. As the ensnared fish attempts to back out of the net, it becomes locked in when the nylon twine bites into the gills under the bony opercula. As the fish twists and pulls its body in an attempt to get free, the twine almost always causes irreversible damage to the delicate tissues of the gills. No matter if the fish is dead when the waterman slips the net over the side of his boat, Maryland law says it cannot be retained. It must be thrown back into the river to decompose and wash up on a beach in a few days.

In prior years the weight of the fish determined if it was a keeper or not. Length replaced weight as the criterion when the authorities came to the realization that a rock's poundage could be lightened considerably if the visceral organs were removed. Watermen and so-called sportsfishermen alike routinely carried as part of their tackle an eighteen-inch stick with decapitated nails driven at right angles through one end. If a fish weighed a couple of pounds in excess of what the law allowed, the "gut puller," as it was called, would be thrust down the mouth into the abdominal cavity, twisted around, and then yanked out with most of the internal organs attached. Apparently someone snitched to the Marine Police or some other official about this unsavory practice, and the law was changed. I found one of these reducing devices in an old abandoned workboat at Tyaskin and have had a lot of fun showing it to visitors and having them try and guess its function.

The stretch of beach just to the north of our property has long been known around the area as one of the best spots for bank fishing for rockfish on this part of the river, apparently because a small gut drains an adjacent acre of marsh into the main channel at this point. As the tides flow in and out of the marsh every twelve hours, large numbers of grass shrimp and minnows are made available to predators cruising offshore. The only practical way to get to this stretch of beach is across our land, and I was warned by the previous owner to be prepared for an influx of uninvited guests each spring. Trespassing and blatant disregard of "Posted" signs by individuals determined to get to the "rock spot," as the area was termed, necessitated my predecessor's having to stretch a chain between two trees to block the road in order to protect himself. I really did not take all of this seriously until one spring day I drove into my backyard and found eight vehicles parked at random. I was stunned when one of the interlopers rushed up to me demanding I hurry to open the house so that he could make a phone call and tend to a pressing call of nature. I glanced around and found my newly seeded grass cut to ribbons by spinning tires and the entire locale littered with discarded cans, bottles, and other trash. Convincing these "sportsmen" to "get the hell out of here" was not an easy task. One group argued that a local bait dealer who sold them several dozen expensive bloodworms gave them directions to my place and told them it was perfectly all right to go in there and do what they wanted. One angler, obviously inebriated, threatened to strike me with the skillet he was using to fry fish over a makeshift fireplace he had constructed near the bulk-

View of river and old Jessie

head. Fortunately, a Wicomico County deputy sheriff lives down the road a bit, and he responded to my call for help and quickly convinced the interlopers it would be in their best interest to leave immediately and not return.

I have never refused anyone permission to park and cross the property if they come to me first and ask. I explain to them I do not enjoy repairing tire ruts or picking up assorted garbage, and if they are willing to respect my rights, they are welcome to use the facilities. Only two groups in the last three years have violated these conditions and had to be "run off," as the river people say.

Jessie and his people have been coming for the last three years and are most welcome. They were one of the few groups to which the previous owner allowed access. Arriving before daybreak, they quietly park and unload their truck. Frequently they erect a tent on the beach for protection against the chilling winds, and the women tend a gasoline stove to insure a constant supply of hot coffee and food. When I venture up the beach to see if they are having any luck, I am always invited to share in anything that may be cooking on the fire. The men see after a dozen or so surf rods with lines attached that have been cast fifty to one hundred yards out into the river. The hooks are always baited with bloodworms, which they buy by the flat (ten dozen). They stay as long as the bloodworms last or until dark, and the last thing they do as they depart the area is to load in the vehicle a sack full of cans, bottles, cartons, and other beach junk they have collected during the day. The beach is always left in a better condition that it was when they arrived. Frequently, I will find by the door of the house a supply of dressed white and black perch or filleted rock they have given me. It would be a lot better world if everyone had the manners and good taste of Jessie and his clan.

Jessie has been bank fishing for rock as long as he can remember and has witnessed the rise and fall in numbers of this fish over the passing years. He recalls the early 1950s were good years as were 1964 and 1967. An outstanding year, 1970 is remembered especially when on one occasion 350 pounds of rock were taken from the beach on their rods and reels. But

Where Have All the Rockfish Gone?

since 1970 everything seems to have gone downhill. Several days during the 1980 season not a single rockfish was taken by the family, and if it had not been for a good catch of perch, they would have returned with nothing to show for their considerable efforts.

Neither Jessie nor his clan know where all the rockfish have gone, but they, like almost everyone else along the river, have some definite ideas. Some people simply explain the rockfish's absence by saying that they have migrated elsewhere, but Jessie thinks this quite unlikely. He puts most of the blame for the rockfish decline on the shoulders of the commercial fishermen and the miles of netting they use to take the spawners as they make their annual runs up and down the river. He is undoubtedly partially correct. One waterman who fishes out of Crisfield is known to place at least *ten miles* of nets in Tangier Sound each spring. A marine policeman once told me he would estimate at least six hundred miles of gill nets were set each year by all the watermen, and this seems to be a fairly accurate figure. One would think it would be impossible for a single fish to escape such a massive snarl of nylon, but, somehow, many of them do.

Jessie also relates he has heard that New Jersey, Virginia, and the Carolinas have few if any laws regulating the capture of the rockfish as they migrate into their coastal waters, and thousands (perhaps millions) of pounds of stripers are taken before they have a chance to enter their ancestral mating grounds to perpetuate the species. This information is basically correct, and it certainly adds to the reasons why the rock is rapidly disappearing from the Nanticoke and other locations.

It has only been since about 1975, however, that people have become concerned about the dramatic decline of this fish. One would think the watermen would be the first to become upset when one of the species they depend on for their uncertain income becomes scarce. Actually they seem to be the last to recognize the decline of a species—be it rockfish, oysters, crabs, or clams—as being important. With a philosophy I find impossible to understand, most watermen shrug the matter off and state the idea that there have always been plenty of rock, oysters, crabs, and clams in the river, and, even though they are a little hard to find this year, they will come back next season. They always have in the past, the watermen say, and there is not a reason in the world why they should not continue to do so in the future.

However, during the last few years, various agencies have become very concerned about the future of the rockfish, and research programs have been instigated in an attempt to find out the important aspects of the facts of life about the rock which are unknown to scientists. Unbelievable as it is, there are numerous glaring gaps in our knowledge of rockfish biology and life history. One would suppose biologists would have long ago worked out the basic bionomics of the most important fish of the Chesapeake Bay system, but this is not the case.

Basically there are eleven know spawning grounds of any importance in the Chesapeake region which are located in the upper parts of the tributaries of the Bay. The fresh, upper reaches of the Nanticoke are the most prolific spawning grounds known, but other rivers such as the Potomac, Choptank, Wicomico, and Manokin are of considerable importance. It is thought that fish bred in these estuaries do not migrate to the Atlantic for a year or more, but this important period of time has not been definitely determined. Leaving the Bay the young rock range far and wide up and down the Atlantic seaboard from the Carolinas to New England. When they become sexually mature, they respond to an overpowering homing instinct incorporated in their genes and return each spring to the area of their own origin to carry out the vital process of reproduction. The forces inspiring and directing these migrations are for the most part unknown. The breeders head for the semifresh waters of the spawning grounds with an inflexible determination. At times a large cow rockfish will frequently bull her way through a gill net blocking the way upstream, leaving the device torn to shreds much to the consternation of the waterman. However, the nets do take untold numbers of these fish as they head for their rendezvous with the males at a place still farther upstream. Many individuals—

Where Have All the Rockfish Gone?

like Jessie and myself—believe netting during the spawning run is a major factor in the decimation of this species' numbers to the point where insufficient young are now being produced to insure the survival of the fish much longer.

Some of the breeders manage to evade the nets and hooks and eventually arrive at the propagating area sometime in early April, and, when conditions are just right, spend the period of dusk to dawn of one night mating. Millions of eggs are laid and hundreds of gallons of male seminal fluid containing the spermatozoa are spread over them. This definitive act of fertilization takes place near the surface of the river, and, as the males and females wriggle and squirm about, the sperm cell passes the chorionic layer of the ovum, crosses the egg membrane, and fuses male deoxyribonucleic acid with female DNA to create a new rockfish.

In the old days in the mating areas along the river, residents said the fish made so much noise during their sexual gyrations it was impossible to get a good night's sleep when spawning occurred. Unfortunately, these "rock fights," as they were called, are now nonexistent or are so subdued it is very unlikely anyone is aware they are taking place—much less lose any sleep because of them. Old-timers also relate that the river had a distinctive—not unpleasant—smell the morning after the night of procreation which reminded them of rockfish. Presumably this aroma resulted from the mixing of the river water with the untold amounts of genital products discharged during the night.

With the act of generation accomplished, the exhausted parents now slowly make their way back down the river to the Bay towards the ocean. Once more they must run the gauntlet of nets and hooks, for the fishermen are still there, and a large percentage are killed and thus prevented from returning next year to breed again. Some river residents with whom I have discussed this matter claim it makes no difference if you take a southbound rockfish (one which is heading back to sea) as they will shortly be dead anyway. They believe that the rock, like the salmon and some other species that migrate from saltwater to fresh to spawn, make a one-way trip. This is not true. Sexually mature adults may return year after year to the same place to reproduce until old age or the nets and hooks finally do away with them.

Millions of fertilized eggs result from the rock fights, and in about three days a larval rockfish will emerge, it is hoped, from each of these zygotes. However, only a few of the fertilized eggs will go through the necessary cleavage stages and subsequent cell divisions required for the development of a new fish. Studies in the last few years have shown only a small portion of the fertilized eggs will hatch, making the tedious and perilous mating runs of the adults hardly worth the time and effort. No one knows why most of the ova will not successfully complete their early embryonic stages, but chemical pollutants are strongly suspected. Eggs taken from old females and fertilized with the milt of captured males have been demonstrated to be almost one hundred percent inviable. Some experts believe these older fish (perhaps they are eight to ten years old) have just lived too long in the presence of chemical residues and have accumulated high concentrations of contraceptives in their bodies. These antifertility compounds may have amassed in the ovaries and spilled over into the developing eggs to such an extent the natural processes of the ova are disrupted. Eggs taken from younger fish and treated in a similar manner have a much higher rate of viability.

The few hundred larval fish resulting from the millions of zygotes are natural prey for many other aquatic species, and of these few hundred, fewer still will be around during the summer months. Those surviving the first year will have attained a length of about seven inches and are now relatively safe from predation. In fact they will have become predators themselves, devouring grass shrimp, aquatic worms, and small fish in great numbers.

By the end of two years the rock will be about a foot long and almost ready to enter puberty. Until fairly recently fish ecologists believed the young rock remained in the general area in which it was spawned for the first year or two of its life and then began its journey out to sea. Newer findings seem to indicate this may not always be the case, and at least some

Where Have All the Rockfish Gone?

young rockfish may head for the open waters much sooner than formerly assumed. As it is essential for biologists to know for sure just what these yearling rocks do during the first year or so of their lives, experiments have been designed to trace their movements. Dr. George Krantz at the Horn Point Environmental Laboratories of the University of Maryland near Cambridge, Maryland, has addressed himself to this and other rockfish problems. Having proposed a scheme to establish an artificial rockfish propagation facility at Horn Point, he would then collect mature males and females from the Chesapeake and artificially remove their sex cells at the time of maturation. Zygotes would be created and the resulting test-tube rockfish would be reared in captivity until they had attained a size that would enable them to have a good chance of survival in natural surroundings. These experimental animals would be marked with a nontoxic, persistent dye visible only with a special light. The tagged rockfish would then be released in various parts of the Bay and its estuaries, and sampling over a period of years would give an indication of the migratory habits of the fish when captured and identified as previously marked. Funds requested for this project in the amount of some six hundred thousand dollars have been requested from various governmental agencies, but have not yet been forthcoming. In addition to data on rockfish movements, valuable information could be gathered concerning other questions about the species, such as what chemicals tend to build up in their bodies, why some rivers and estuaries, and not others, are devoid of them, and at what life stages are they initially susceptible to various types of mortality.

Researchers have observed for some years that there seems to be an unusually high death rate in juvenile rockfish, and members of the U. S. Fish and Wildlife Service, who have studied immature fish from the Hudson River in New York, and the Potomac and Nanticoke rivers in Maryland, think they know the reason. Many young fish examined were found to have weakened backbones, a condition linked to the presence in their bodies of two common pollutants of rivers, selenium and arsenic. The former element is common in the ash emitted by coal-burning power plants, and the latter is a common waste product of many industries that may dump this material into the aquatic environment. Apparently these substances interfere with the deposition of the normal amount of proteinaceous material necessary for a healthy skeleton. This weakened spine affects the overall strength of the animal, putting it at a distinct disadvantage. It finds itself unable to compete with its rivals for the necessities of life and, through the process of natural selection, is eliminated. The basic law of nature is that only the strongest and best adapted will survive to perpetuate the species.

Other studies have been made that indicate certain pesticides and fertilizers washing into the habitat of the rockfish act to extract vitamin D from its bones and weaken the overall skeleton to a significant extent. If, indeed, the young rock remains in its spawning area for one or two years or more, he is certainly more apt to absorb debilitating amounts of selenium, arsenic, pesticides, and fertilizers, than if he leaves for the relatively unpolluted open seas soon after he is born. This is why knowledge of what young stripers do is essential to the understanding of why their numbers are growing fewer and fewer with each passing year.

Other researchers have suggested that pollutants have not harmed the rockfish directly as much as they have harmed it indirectly through the food chain. There seems to be no question concerning the harmful effects of certain pollutants on the all important plankton. These microorganisms comprise the vast majority of the food of the larval rockfish from the time they emerge from the egg until they have grown to the fingerling size. It could be that the plankton serves as a vehicle to transfer effectively the contaminants from the water to the body of the fish. This has proven to be the case with the buildup of high pesticide residues in other species studied. Other specialists have suggested that the pollutants are responsible for a change in the overall ecology, which in turn adversely affects the production of plankton. And so it goes on, as answers to the pressing problem are sought by scientists and laymen alike.

Where Have All the Rockfish Gone?

The Maryland Natural Resources Department is feeling the pressure from various interest groups to do something about the situation, even if it is of a "band-aid" nature. For example, during 1980 the Department attempted to push a bill through the legislature prohibiting the use of gill nets by owners of waterfront property without a commercial fishing license. The practice of stringing a net of up to one hundred yards in front of their property for a small fee is a time-honored tradition enjoyed by the residents of the Nanticoke, and they were not about to give up this "inherent right" without a fight. In spite of figures submitted by the experts in Annapolis proving that thousands of pounds of rockfish could be saved if this measure was enacted into law, it was blocked by the delegates from the Eastern Shore who obviously know which side of the broiled rockfish the butter is on. Undaunted, the Natural Resources lobby pushes on with proposed conservation measures that seem to change so often it is difficult if not impossible to keep abreast of them. Measures to prohibit fishing of any type for rock during the spring and early summer months, a restriction on the mesh size of the nets, and new size limits are examples of their recent attempts to halt the decline.

The restriction on the use of mesh size was designed to prevent the large—supposedly superbreeders—from becoming entrapped. Yet, as we noted previously, the huge, older fish are almost completely sterile, and if any successful reproduction is to take place, it will be accomplished by the younger and smaller fish which are, at least to some extent, capable of producing fertile gametes. These fecund smaller fish are the very ones most likely to be taken in the smaller meshed nets that are now required.

It seems to me that if we have the technical knowledge necessary to put a man on the moon and to create new forms of life by gene splicing, we should be able to figure out where all the rockfish have gone and to do something about it.

Last spring I rigged a long spinning rod with bloodworms and cast them far out into the river in front of the house. I set the rod in a sand spike and went about tending to various chores about the yard. A few minutes later I noticed the rod bending under a strain from the line, and I immediately grabbed it and reared backwards. I had hooked a rockfish that required about ten minutes to guide onto the beach. I could see it was an egg-laden female that would tip the scales at about fifteen pounds—a legal fish. I immediately thought about how many pounds of delicious fillets could be sliced from this beauty, but even as these thoughts were passing through my mind I was in the act of disengaging the hook and carefully guiding the fish out into deeper water. I missed many a fine meal by this action, but somehow I felt good when I watched the expectant mother gracefully navigate her way out from the beach and head upstream, and I could only hope that a few of the ripening eggs in her abdomen would be fertile.

9

Of Turtles and Snakes

ONE OF THE FIRST THINGS I did when I came to the river was to acquire a couple of wire crab pots or traps and place them in the shallow water in front of the cottage. As it was early in May, the blue crabs were just beginning to return to our waters, and I did not expect to harvest a bushel of the crustaceans in a twenty-four hour period. I did expect to catch more than the five or six undersized specimens and the two turtles I discovered in my pots the next day. In disappointment I dumped the crabs back into the water and flung the turtles after them. One of my neighbors happened to be walking along the beach at that time and witnessed my actions. "Are you crazy?" he yelled at me, and added, "Them is diamondbacks you just chunked away."

It suddenly occurred to me that he was talking about the diamondback terrapin—the gourmet's delight, the mascot of the University of Maryland, and, a generation ago, the most valuable reptile in the United States, if not in the entire world. My only acquaintance with this reptile had been a few years previously when I ordered a bowl of terrapin chowder in a quaint restaurant in the village of Oxford, Maryland. I was more impressed with the price ($1.75) than I was with the chowder or the tiny bits of "terrapin" floating around in it. My neighbor responded to my lack of interest in the animal by giving me a short course in "terrapinology" which stimulated my curiosity and taste buds.

No member of the turtle tribe has ever enjoyed the fame and popularity the diamondback terrapin experienced during the period in our history that extended from the gay nineties to the roaring twenties. (Incidentally, a terrapin is usually defined as an edible turtle found in water; a tortoise is usually a turtle found on land; and a turtle may be either one or the other.) It has been said that no champagne supper was complete without diamondback terrapin among the delicacies to be consumed. This terrapin is called the diamondback because the concentric rings within the plates of its shell resemble diamonds. But, in addition, diamondback is also an appropriate name for these beasts as during the period of their extreme popularity they were almost worth their weight in diamonds. During the late twenties a dozen diamondbacks measuring from seven to eight inches along the bottom shell and weighing about two pounds each sold for from ninety to one hundred dollars a dozen on the hoof. Appropriately, it is reported that Diamond Jim Brady was fonder of terrapin than he was of oysters, and I imagine he hosted many a dinner party for celebrities of the day that dented the terrapin population. One gourmet of the era referred to the terrapin as, "the scintillating gem in the dietary of the elite," and further remarked, "these animals are not intended for the vulgar palate."

Terrapin did not always evoke such flattering remarks. Extremely plentiful in early America, it formed a significant portion of the colonists' diet at certain times of the year. Martha Washington's cookbook included a recipe for terrapin as did other colonial publications. These animals were so plentiful then that plantation owners literally stuffed their slaves with the succulent meat. However, masters overdid a good thing, and in Tidewater Maryland some of the slaves refused to perform their chores in an attempt to gain relief from a diet too rich in terrapin. The legislature responded to the rebellious slaves by enacting a law that forbade a slave owner from feeding terrapin to his charges more frequently than once a week.

Sometime between then and the late nineteenth century things changed for the better, or the worse, depending on the

Diamondback terrapin

point of view (man's or turtle's). By the late 1800s the diamondback terrapin was the most popular item on the menu of the discriminating diner. No one seems to know how or why this fad for diamondbacks developed. Maybe it had to do with the overall flamboyant, extravagant life-style of that time, or, perhaps, someone rediscovered the gustatory delights of a creature with which most people had grown bored. In any event, the demand by the sophisticated epicure far exceeded the supply. In 1891, eighty-nine thousand pounds of terrapin were reported taken in Maryland alone and sold for an average price of 25 cents a pound. No species can withstand such abnormal pressures on its numbers for long. By 1920 only 823 pounds from Maryland were sold at a price of $1.22 per pound. Based on an average weight of 2 pounds each, these figures translate into a catch of only 410 turtles, each of which was worth about $2.50. During the remainder of the twenties the reptiles became scarcer and scarcer, and by 1929 they were fetching the aforementioned astronomical figure of $90 to $100 a dozen.

The turning point in the life of the diamondback was in 1929 when conservation laws were passed by several states in an attempt to save it from extinction. The taking of diamondbacks at any time was forbidden in a few states, but other states passed laws that were not quite so stringent. In Maryland a closed season during the mating period was declared, and the sale or even possession of these animals in that period was unlawful. A minimum size of five inches was established, and it was illegal to molest, possess, or sell the eggs at any time. These laws were passed in the nick of time because if the wanton destruction of their numbers had been allowed to continue the terrapin surely in a few years would have followed the path of the passenger pigeon to extermination.

By 1935 the population of terrapins had rebuilt itself to such an extent that five thousand eight hundred pounds of Maryland terps were captured for the market. But, even today, with conservation laws on the books and enforced for fifty years, the populations of terrapins are only a small fraction of their previous sizes. Now old-timers along the river say that where you may see two or three terps in a given area in a day's time, years ago a hundred or more would have been present.

Fortunately for the continued existence of the diamondback the craze for Terrapin à la Maryland went out of style almost overnight in the early 1930s. Several reasons have been suggested for this sudden lack of demand. One explanation is that Prohibition made the supply of alcoholic spirits used in preparing the turtle, and always imbibed while eating it, very difficult to obtain. A more logical reason seems to me to be that the florid style of living and entertaining of an earlier era was greatly curtailed by the economic depression of the country during the 1930s. With people standing in bread lines and selling apples on street corners, it would seem almost sinful to tease one's palate with a bit of reptile costing as much as many people earned in a week.

The deficiency in numbers of the terrapin during the years of its peak popularity led the now defunct United States Bureau of Fisheries into the business of artificially propagating the species. The main terrapin ranch was established at Beaufort, North Carolina, and endured for the better part of forty years. Enterprising individuals with an eye for easy money also got into the act, and numerous private terrapin farms sprang up along the eastern seaboard. The interest in these animals was so great a generation ago that some of the turtle nurseries were opened up for tourists. One such turtle factory is reported to have featured a trained terrapin that simulated playing an old-fashioned player piano to the delight of gawking spectators. With the decline of the terrapin-meat fad, however, the private propagators gradually went out of business, and even the federal government eventually terminated its program which might appear to be a minor miracle in itself.

Today the diamondback is fairly plentiful along the Nanticoke and other rivers of the area. The range of this terrapin is extensive, as it is found in the coastal marshes from New England to Mexico. Apparently this is the only type of turtle limited to this kind of distribution pattern.

The terrapins mate only in the water during the spring months. The male rides on the back of the female during a rather lengthy mating process, and after fertilization is accomplished, the female comes ashore to seek a suitable spot for the deposition of her eggs. Apparently she is quite finicky and may roam around seeking just the right spot to build her nest. When she finally decides on a spot which may be in a sandy ridge bordering a salt marsh, she invariably empties the contents of her bladder on the spot to facilitate digging. With her sharp claws she scoops out a nest which is roughly triangular in shape, four to eight inches deep. Four to twelve eggs are deposited in the nest and covered with various types of detritus. With all maternal instincts apparently exhausted, the female terrapin leaves the nest to the dangers of nature, relying on the sun to incubate the eggs. Approximately ninety days of incubation are required before the eggs hatch. A female terrapin becomes sexually mature during her fifth or sixth year of life and continues to be fertile for an estimated twenty-five years. The life of a diamondback is suspected to be in excess of forty years.

Several clutches of eggs may be deposited by the same female during a single season. One investigator reported a female he observed for an entire season laid five separate clutches. The adult turtles have few enemies other than man, but the eggs are prized as food by muskrats, skunks, hogs, and crows. Therefore, it is well for the survival of the species that the female is so prolific in her egg-producing abilities as a large percentage of eggs she lays are destroyed by predators.

During the warmer months of the year the diamondbacks bask in the sun along the edge of the marsh, or they may float in the river for hours at a time. They drift along in the water in a vertical posture with just the tip of the snout protruding. The broad, webbed feet are used to tread water, after a fashion, and a terrapin forages for food as it moves along ever so slowly. It is not a fastidious feeder and will readily devour snails and other small mollusks, crustaceans, annelid worms, and bits of vegetation. On land they have been frequently observed to nibble on marsh grass and undoubtedly will scavenge the bodies of animals either in the water or on the shore. At night they retreat to the seclusion of a gut or drainage ditch

Of Turtles and Snakes

where they burrow into the mud to rest from the day's activities.

With the coming of fall terrapins enter into a state of almost complete hibernation, buried in the mud at the bottom of the river. A hormone is produced that acts to constrict the blood vessels supplying the organs of the body, and as a result the metabolism of the animal is reduced to practically zero. What little energy is needed to keep the heart beating a few times a minute is provided by the physiological process known as glycolysis. In this operation no oxygen is required to respire stored foodstuffs, and none of the poisonous carbon dioxide is produced as is the case when oxygen is used to burn the fuel of the body. Terrapins may remain in this state of suspended animation for months at a time, or if the water happens to warm considerably, they can become active again before re-burrowing into the bottom as the temperature declines.

It is during the hibernation period that hunters of the elusive diamondback make their largest hauls. A special rig known as a terrapin dredge is pulled and scraped along the bottom to capture these sleeping beauties. I have heard persistent rumors to the effect that there are a couple of small, preferred hibernating locales situated somewhere at the mouths of the Wicomico and Nanticoke rivers. These areas are supposed to be known only to a few watermen of the area, and the exact locations of the terrapin gathering places are a closely guarded secret. Knowledge about the sites where the treasure is buried has supposedly been handed down in only a couple of families for generations. As the story goes, the few persons possessing this top secret information are able to "drudge" up hundreds of terps during the winter season from those areas, reported to be no more than two acres in size.

For the rest of us along the river who are not privy to this sacred information, about the only way to acquire a diamondback during the hibernating season is to probe in the mud on the bottoms of the shallow tributaries to the river. With a little practice one can soon recognize the feel of the shell of a turtle with the pointed end of his staff and quickly distinguish this sensation with that given by a rock or other inanimate object.

During the summer months the diamondback can be taken in an ordinary wire crab pot which has had the entrance funnels enlarged sufficiently to admit the curious and/or hungry reptile. If terrapins are trapped in this manner, it is a good idea to check the pots every six to ten hours. Turtles with their metabolic processes going full blast require oxygen to sustain life. Though they can remain submerged for a few hours they will drown if oxygen is denied them for an extended time.

Some serious trappers use fyke or pound nets to capture terrapin along with various species of fish. These types of nets are designed to make it easy for a turtle to enter but almost impossible for him to navigate his way out. It is imperative the nets be checked frequently, as dead terrapin has no market value. Terrapins that have been dead for only an hour or so, however, should not be discarded as the quality of the meat is not affected by a short period of lifelessness. Terrapin fanciers should understand that a frozen turtle has a taste similar to shoe leather. It is axiomatic that when one orders turtle in a restaurant he should always determine that the meat has been freshly killed and not frozen for weeks on end.

One individual who routinely gathers both terrapins and snapping turtles for the commercial market told me that a five-inch terrapin brought $2.75 on the Philadelphia market during 1979. Those in the six- to eight-inch class fetched the handsome price of $5.00. With these inflated prices it seems the cycle is reversing itself again, and terrapin meat is making a comeback. As the turtle capital of the United States, Philadelphia receives almost all of the reptiles caught and shipped commercially. Baltimore receives a small percentage of the catch. A friend of mine visited a famous seafood restaurant in the City of Brotherly Love last year and noticed Terrapin à la Maryland was listed as an entree on the menu for the outrageous price of $13.00. In contrast, the price of a whole Maine lobster was quoted as $7.50.

Most residents of the Nanticoke who partake of terrapin—there are many who refuse to touch it—agree the animal should be killed by placing it in boiling water and left there to cook for the better part of an hour. When a fork can easily

Of Turtles and Snakes

penetrate the skin, the turtle is ready for cleaning. The bridges between the upper and lower portions of the shell are severed with a pair of tin snips, and the internal organs are removed and discarded. The liver, however, is retained after the gall bladder has been carefully dissected out. The meat is then scraped from the shells and the neck, and the legs are skinned and defleshed. If one is careful and little meat is wasted, a two-pound terrapin will yield about a quarter of a pound of boneless tenderloin. There are endless recipes for preparing the meat, but I have found it to be delicious when sauteed in butter or margarine with chopped onions and diced boiled potatoes added. Even if the champagne is not available, one cannot help but feel a little like the bon vivant of bygone days when he dines on Nanticoke diamondback.

Snappers

Snapping Turtle à la Maryland does not sound nearly as appetizing as Terrapin à la Maryland, but there are numerous individuals who will argue that the flesh of the snapper is every bit as delicious as that of the diamondback. I have heard rumors to the effect that eating establishments frequently substitute snapping turtle meat for the more expensive terrapin, and the diner is none the wiser. This large reptile (some may reach three feet in length and weigh forty pounds) enjoys a much wider distribution than the terrapin as it is found in almost all rivers and large bodies of water east of the Rockies. The snapper is an aquatic animal and dines mainly on fish. To entice its supper into its lethal mouth, the snapper waves in the current several pale pink, filamentous processes attached to its tongue. The unsuspecting fish finds out too late that the wormlike filaments are just one of the snapper's evolutionary adaptations.

The female snapping turtles wander away from the water to find a suitable nesting area and sometimes turn up in a backyard or are almost run over as they amble down a country road. I recall going on an errand with a friend who lives on the river when suddenly he slammed on the brakes of his pickup truck. "There goes the best tasting meat a man can put in his mouth," he exclaimed as he pointed to a huge snapper disappearing into the weeds at the side of the road. With that he jumped out of the truck, and, without a moment's hesitation, deftly grabbed the monster by the tail and flung it into the back of the truck. I was sure the animal must be infuriated by this ungracious treatment; and when I poked at it with an old broom handle, the head darted forward with unbelievable speed. The massive jaws cut the handle in two as neatly as if it had been struck by a sharp ax. I have no doubt that those sharp jaws can meet with such force that fingers or even hands of the unwary could be nipped off in the twinkling of an eye. However, I have never had a report of anyone along the Nanticoke being bitten by a snapper, though I feel sure this has occurred.

I asked my friend how in the world he was going to kill and dress this huge beast. He explained that the proper way to dispatch a snapper is first to strap it to a tree upside down. When thusly secured, the head is grasped firmly with a pair of very long-nosed pliers. With a tug the head is pulled from the protection of the shell and stretched out along the tree, and the animal is quickly decapitated with a keen ax. The headless beast is then left lashed to the tree to drain for a half hour or so. During this period my friend advised that the amputated head be picked up with a pair of pliers and buried immediately. He explained this was necessary as the severed head remained capable of chomping down on anything that touched it until the sun went down. It is possible for such an event to take place—even after sundown—because if the jaws were open when the head was removed, one could stimulate a severed nerve ending by handling the head, and the terrible jaws could snap shut by reflex action. This probably happened to someone at one time or another and gave rise to the legend that the head lived on after it was separated from the body and sought to revenge itself.

Snapping turtles are not as easy to come by along the river as the terrapins are. I have it on good account, however, that there are three individuals in the area who actively pursue the snapper for the Philadelphia market. The wholesalers there

Of Turtles and Snakes

were anxious to pay fifty cents a pound for all the live snappers they could get during 1979. Since snapping turtles are voracious predators, the environment can support only a limited number of them, and the trapper never encounters more than a few fully grown snappers in a given area. There are just not enough prey animals to go around to satisfy their unlimited appetite. The best way to catch a snapper is with a hook and line baited with salt-cured eel meat. This method is illegal under Maryland law, but it is still the most common way snappers are taken. Currently there is no closed season on the snapping turtle in Maryland, but, in addition to the prohibition on taking them with hook and line, it is also unlawful to catch a snapping turtle with any device capable of piercing the skin. Of course, this outlaws the bow and arrow, gig, spear of any kind, and firearm.

The snapping turtle is a sinister looking beast who is just as mean and dangerous as he appears. His overgrown cousin, the alligator snapper of the southern states, is even more ill-tempered and dangerous because of his inflated size. Routinely this monster tips the scales at 150 pounds, and there is one on record that almost broke the scales at 219 pounds. There are several documented accounts of the alligator snapper attacking man, and a number of individuals have lost fingers, hands, and the lower portions of their legs when crossing the path of this huge beast. Both of these turtles also take a toll of unsuspecting waterfowl, and for this reason they are almost universally detested by waterfowl hunters.

As mentioned, the meat of the snapping turtle is placed in a class with the terrapin from a taste standpoint. One snapper epicure along the river always delays the execution of any turtle he acquires for about a week. His snappers pass a stay on death row in a swill barrel filled with food scraps on which they are supposed to fatten themselves up for the eventual kill. It has not been my pleasure so far to dine on snapping turtle (unless I have been duped by some restaurateur who served it under the name of terrapin), but I will jump at the chance if the opportunity presents itself. However, this occasion will have to be provided by someone else who has captured the beast and done the necessary preparation. I am not about to wrestle with so formidable a creature just to obtain a few pounds of turtle tenderloin—I value my fingers too much.

Copperheads, Cottonmouths, and Other Snakes

The only two poisonous snakes found in Maryland are the timber rattlesnake and the copperhead. The rattler is confined to the far western counties, while the copperhead is a resident of all counties of the state. But, try to tell ninety percent of the Eastern Shoremen that the copperhead is the sole venomous snake in the Tidewater region. Most shore inhabitants are absolutely positive the cottonmouth moccasin is abundant throughout the area and poses a definite threat to their lives. The truth of the matter is the cottonmouth moccasin is found naturally occurring only as far north in the United States as the region of Norfolk, Virginia. Some years ago a moccasin was killed in the wild on the lower Eastern Shore, but after investigation of the occurrence, herpetologists believe the reptile was released there after capture elsewhere.

The northern water snake is the species most commonly confused with the deadly cottonmouth. The water snake is quite abundant and can be found in residence in virtually every swamp, marsh, bog, stream, or lake margin from southern Quebec to North Carolina. This is a large serpent which averages about three feet in length with the record being four feet three inches. It is a ferocious beast that would rather fight than run when encountered on land. When disturbed in its terrestrial setting it will frequently coil and vibrate the tip of its tail producing a sound not dissimilar to that made by a rattlesnake. In the coiled posture the enraged water snake will often rear back its head and open its mouth in a threatening gesture. It is certainly true the inside of that gaping mouth does appear "cottony," and it is not surprising that the startled human may mistake it for a true cottonmouth if this is the only identification criterion he uses.

The true cottonmouth is a member of the pit viper family of snakes and is closely related to the copperhead and rat-

Northern water snake

tlesnake. The pit vipers are so named because there is a sensory pit on each side of the head between the opening of the external nare and the eye. The pupil of the eye of the pit viper is elliptical, and the body is stout and chunky with the tail being rather short. Its general coloration is olive brown with numerous dark crossbars. All of these characteristics of the cottonmouth moccasin can be applied loosely to the northern water snake with two exceptions. The harmless water snake has no pit on each side of its head, and the pupil of the eye is round in shape. In areas where the two species occur together, the only sure way to make an accurate identification is to check for the pits and the shape of the pupil. Of course, the moccasin has two long, hollow fangs by which the venom is injected. These fangs are not present in the water snake. Getting close enough to a snake of questionable identity to determine if it has a round or oval shaped pupil, a pit, or fangs is not recommended for the average person. It is best to avoid contact with water snakes in areas where their distribution range overlaps with that of the cottonmouth.

One characteristic the old-time residents of the Nanticoke (and many present day residents, as a matter of fact) have in common is their fear of all snakes, and the northern water snake seems to create a deeper sense of apprehension and alarm than any of the others. When you inquire of various people if they personally know of anyone who was bitten and subsequently envenomized by a water snake, the answer is always "no." You are also told the correct name for the reptile is moccasin or cottonmouth and not water snake. After this lesson in species nomenclature is emphasized, it will be pointed out that "everyone" knows the moccasin or cottonmouth is deadly poisonous, and since they are so abundant in the region they surely must have bitten and killed many people. It is just a matter of not knowing directly of anyone in this category.

I was fishing once in Jack's Creek with a retired waterman from Wetipquin, and as evening approached several large water snakes could be seen swimming out of the marsh in search of their evening meal. My companion became extreme-

Of Turtles and Snakes

ly nervous and insisted it was time to, "Haul ass outta here." He was afraid the snakes would board the boat and attack us.

Another story is told of a Bivalve resident who hooked a water snake on his rod and reel while fishing in Newfoundland Gut. As he reeled in his catch, he mistook the snake for an eel and swung the infuriated reptile into the boat. Some say his error in judgment may be attributed to the fact he always included a pint jar of moonshine in his fishing gear. Be that as it may, the Bivalver quickly realized he was involved in a case of mistaken identity when the snake threw the hook from its mouth the moment it hit the deck of the boat and came up as if it were as mad as hell and ready to fight. In a state of terror, the fisherman leaped out of the boat into the waist-deep water of the gut and slogged through the muddy bottom to the shore. As the boat was not anchored when the snake came aboard, it proceeded to drift out of the gut into the river with the ebbing tide. After some time had passed the terrified angler was able to signal a passing craft which picked him up off the beach. His skiff had drifted a quarter of a mile down the river, and he and his rescuer approached it with great caution. When the derelict boat was overtaken, it was finally determined the snake was no longer aboard, having been able to crawl over the gunwale and slip back into the water. It is said the man quickly sold the boat after this incident because he could never feel comfortable in it again.

As previously noted, the copperhead is found in all parts of the Eastern Shore and is a true pit viper capable of inflicting a dangerous bite that infrequently is fatal. For some reason the copperhead is less feared by the residents of the Nanticoke than the water snake. This is probably because the copperhead is not nearly as numerous as the aquatic species, and it is seldom that a copperhead is encountered in the area. However, it is more abundant than its infrequent sighting would indicate. Residents know it is a poisonous snake to be avoided at all costs and killed if possible.

Herpetologists agree the copperhead is the least dangerous of the pit vipers found in the United States. I mentioned this fact one day when I was talking with a group of local retirees who routinely sit around and swap stories for hours under the shelter in the small park adjacent to the boat-launching ramp at Nanticoke. I was at once corrected by one of these ancient mariners the others refer to as the "Talking Man." Bluntly, the Talking Man told me I didn't know "what in the hell I was talking about." He proceeded to tell me about a personal experience that happened around 1925 near Tyaskin. He and a group of men were digging potatoes in an overgrown field when one of them was bitten on the arm by a copperhead. The others leaped to his assistance, immediately poured almost a pint of whiskey into him, and unceremoniously loaded the unfortunate victim into the back of a horse-drawn wagon for a trip to the hospital in Salisbury. The eighteen-mile trip required about two hours. When the group pulled into the receiving area of the medical facility, the potato digger was in a bad way. Despite medical attention, his condition deteriorated, and he died several hours later. The Talking Man emphasized that I should never again tell anyone that the copperhead was not a very dangerous snake.

Of course, the administration of stimulants of any type to a snakebite victim is the worst thing you can do. As the circulation is increased, the venom quickly spreads to the vital parts of the body. The snake bitten individual should be kept as quiet as possible while the ligature-cut-suck method of first aid is administered. Competent medical attention should be sought immediately, and an anti-venom which is effective against the venom of all pit vipers should be administered.

Two other species of common snakes of the area are also feared by a majority of the inhabitants. The black snake, or blue racer, is one of these and grows to a length of several feet. Some people believe the black snake is venomous, and others are convinced it will "run you," wrap around your body, and slowly squeeze you to death. The black snake is guilty of none of these actions, and is, in fact, a beneficial animal as it destroys large numbers of rodents and other pests.

The hognose snake is the other species that strikes fear into the hearts of many residents. The tip of the nose of this snake

Of Turtles and Snakes

is turned upwards in a menacing sort of way, and the head is broadly triangular like a copperhead's. The hognose is frequently called the spread adder, hissing adder, or death adder. Black snakes and water snakes will bite if you pick them up and the small, sharp teeth may even bring a trickle of blood, but it is doubtful if a hognose snake has ever bitten a human being even when roughly handled. The hognose feeds almost exclusively on toads and frogs, and when confronted by an enemy it becomes an extraordinary bluffer. If danger is sensed, the snake will open its mouth widely and emit a loud hissing sound in an effort to frighten its attacker. It is a common belief that the snake spews forth a toxic venom as it hisses and snorts, but this is not true. If the original bluffing act does not discourage the tormentor, the hognose will flop over on its backside and play dead for an extended period of time. It will usually inflate its body with air during this act to enhance its defensive posture. The hissing adder vaguely resembles a copperhead, and I guess this is the primary reason folklore has placed it in the category of a dangerous reptile. As a result it is killed by most people on sight.

Residents along the Nanticoke take their snakes seriously, and one thing I have learned over the past few years is not to try and correct any belief Nanticokians may have concerning these creatures. When it comes to fish, oysters, waterfowl, and other forms of life in the area they may have open minds and be receptive to new ideas and scientific fact. However, where the legless reptiles are involved, there is absolutely no room for discussion. Their beliefs are so deeply ingrained, it is useless to attempt to pass on any form of the truth.

10

The Return of the Erne

THE SHRILL, PIERCING SCREAM made by the bald eagle as it dove to the attack startled me. I glanced upwards in time to see the eagle make a pass at an osprey carrying something in its talons. The smaller bird disengaged its claws from the object, and it fell toward the surface of the river with the eagle in hot pursuit. The eagle grabbed it just before it hit the water and released it almost immediately when he determined it was a stick and not a fish. The disgruntled bully then flew back toward the tree line of the river after emitting another scream in an attempt to intimidate the osprey further.

The osprey (sometimes called the erne, ern, sea eagle, or fish hawk) did not seem to be overly perturbed or particularly impressed by this attack by the overpowering symbol of the United States. He had probably been subjected to such degradations many times in the past. The erne had learned long ago that when a "true" eagle demanded something he possessed, the prudent thing to do was to comply immediately and entertain no thought of resistance.

The scenario just described took place during the summer of 1979 on the Nanticoke River just north of where it feeds into Tangier Sound and the Chesapeake Bay on the Eastern Shore of Maryland. It is highly unlikely that such a drama would have been observed ten years ago as the numbers of the magnificent participants had fallen to dangerously low levels.

The Return of the Erne

Both of these species, however, have managed to make a comeback in the region of the Chesapeake during the 1970s, although the bald eagle has a long way to go before it can be considered safe from the threat of extinction.

Not so with the osprey! During the last few years its numbers have multiplied dramatically, and today it is considered a rather common species throughout this region. The decline in numbers of ospreys (and eagles as well) during the fifties and sixties can be directly attributed to two causes: widespread pollution of the environment by persistent chlorinated-hydrocarbon type insecticides and destruction of habitat and nesting areas.

For almost twenty-five years after the end of World War II, DDT and other long lasting chlorinated compounds were broadcast on the environment in untold quantities. There is no question that tremendous benefits in the fields of agriculture and disease control accrued to mankind, but there were also some important negative aspects of the saturation of the landscape with the "Magic Bullets" as the compounds were popularly called. A large number of wildlife species were hard hit. Predator birds, such as the osprey and the eagle, were indirectly loaded up with these poisons when they fed on their normal diet of fish which had previously amassed dangerous doses of the toxins from the animals they used for food. Many ospreys were killed outright when they ingested a lethal concentration, but others were subjected to a more subtle type of annihilation.

When ingested in sublethal concentrations, DDT built up in the birds' tissues and adversely affected the reproductive systems. Researchers noted that ospreys with a high level of DDT in their bodies laid eggs that were frequently infertile. In addition, the chemical was found to cause a malfunction of the shell gland so that the covering placed around the egg at the time of laying was abnormally thin. Eggs in museum collections that were collected prior to the advent of DDT were found to have significantly thicker shells than those laid after the use of DDT became widespread. A large portion of these emaciated eggs (even though they might be fertile when laid) were accidentally crushed in the nest under the weight of the parents during the month long incubation period.

Fortunately for wildlife (and nonwildlife too), DDT and the other persistent insecticidal compounds were banned for general use in the United States in 1969. Since that time the increase in the populations of birds of prey has been nothing short of miraculous.

The destruction of nesting sites by man was another major factor that resulted in the osprey being placed in the "Threatened" category of the official Endangered Species List. Destruction of their habitat was accelerated by the skyrocketing value of land on or near the water and the subsequent commercial development of what had been more or less wilderness territory. Many nesting sites were destroyed, however, by bureaucratic regulations and just plain ignorance.

The osprey builds its aerie of sticks and other materials in a variety of situations. One favored building site is within the framework of the many lighted buoys that mark the channels in coastal waters. The United States Coast Guard is charged with the placing, operation, and maintenance of these navigational aids, and until a few years ago, took a dim view of any animal having any contact whatsoever with them. Fishermen were prohibited from anchoring their boats to them, and ospreys were forbidden to build a nest on a "light." Consequently, Coast Guard maintenance crews systematically dismantled and destroyed every nest they found illegally reposing on government property. Some osprey experts believe this wanton destruction of hundreds (if not thousands) of potential ernes was as important as the pesticides in reducing their numbers to the danger point.

In the early 70s, the Fish and Wildlife Service of the Department of the Interior explained the facts of osprey life to the protectors of our waters, and the routine destruction of the nests was discontinued. The present policy of the Coast Guard in this regard is not to disturb any osprey nest built on an aid to navigation unless it is interfering with the proper function of the device. The nest shall be rearranged if possible

Return of the erne

so that the beacon is not obscured and shall not be completely dismantled unless absolutely necessary. In the end human safety is placed above osprey reproduction, and well it should be.

A few osprey are shot each year by a small number of uninformed "sportsmen" who say they believe the piscatorial diet of the fish hawk is responsible for their own lack of success with the rod and reel. Fortunately there are not too many of these individuals prowling our rivers and marshes, and the loss of ospreys to the shotgun or rifle is considered to be minimal.

The change in Coast Guard policy, coupled with the prohibition of DDT, has resulted in an almost overnight increase in erne numbers. In fact, their numbers have swollen to the point where they were removed from the threatened category in 1976. In the 60s or early 70s one could spend an entire day on the water and, if lucky, observe one osprey. Now, a day on the water will yield many sightings of this proud and graceful bird. I was extremely pleased to note that ninety percent of the buoys I inspected along a fifteen mile stretch of the Nanticoke during the spring of 1979 harbored an active osprey nest.

The osprey is truly the Izaak Walton among the birds of prey. It is always a spine-tingling thrill to watch a soaring osprey suddenly fold back its wings and go into a power dive straight toward the water. With a resounding splash the erne will hit the water feet first and disappear for a moment while he impales a fish with his sharp talons. He soon reappears, orients himself on the surface, and then flies away with his finny victim held foremost. If the fish is small, he will devour it while on the wing, but if the erne has been lucky enough to attach onto a lunker, it will be borne away to a secluded spot where it can be eaten in a leisurely manner—that is, if there is no bald eagle in the vicinity to force the surrender of the victuals.

Some individuals have wondered how the fish hawk can dive into the water, catch his supper, reemerge, and fly off immediately without his feathers even being ruffled. Ospreys,

The Return of the Erne

and other birds that routinely dive beneath the surface, have a special oil producing gland at the base of the tail. This waterproofing material is spread to all parts of the body by their frequent preening.

On one occasion, I watched an erne dive beneath the surface and fail to reappear. The only conclusion that I could reach was that he had used poor judgment and sunk his steel like hooks into a fish too large to handle, and, being unable to disengage his talons, was dragged down into the depths and eventually drowned. In this case the fish must have been quite large (if this is indeed the reason the bird did not resurface) as the erne itself is a large, conspicuous bird almost two feet in length. Its wingspan may measure close to five feet or more. The upper parts are dark mottled brown in color with the underside and head being largely white. In fact, the osprey is the only large bird of prey in North America that is clear white underneath. The head has a dark stripe that reaches from the beak laterally to the neck region, and there is a brown patch on the leading edge of each wing. Ospreys may be recognized on the wing at a distance by the marked bow or kink in their wings. Eagles, buzzards, and other birds that might be confused with an osprey at a distance, soar with flat wings. The beak is strongly recurved, and the talons are long and needle-pointed. The erne may be found over rivers, lakes, and other bodies of water anywhere in North America, but are more numerous in coastal regions.

Ospreys have a strong homing instinct and frequently will return to the same nesting site year after year. If the old nest has been destroyed, they will rebuild it, and if it remains, they will remodel and enlarge it each year with additional sticks, pieces of driftwood, cornstalks, seaweed, and other materials that might be available. Such an aerie that is added-on-to each year can become quite large and bulky, and, if located on a buoy, could obstruct the light that guides the nocturnal mariner. The nests may be constructed in the tops of tall dead trees, on telephone poles, or even on the ground, but always in close proximity to the water and the fish it contains. There are usually three eggs in a clutch which is deposited in mid-spring. Twenty-six days of incubation are usually required for hatching, and the fledglings remain in the nest for approximately two months. The nest is guarded at all times by one or both parents.

The osprey has no inherent enemies other than man (the overbearing eagle cannot be considered a true enemy). Potential predators are discouraged from approaching the nest by frantic gyrations of the parents while they emit their eerie, squealing whistle. The scimitarlike beak and the razor-sharp talons must surely discourage strongly any animal who would have foolish designs on the eggs or the fledglings. There is also strength in numbers as the ospreys often nest in small colonies, and, in the breeding season, the air is full of the clamor of the young and the whistles of the adults.

Except when the aerie is approached, the osprey appears to have an easy going personality, and other unobtrusive birds have been known to take advantage of its good disposition. Once in March, I was reposing on the banks of the Nanticoke watching a pair of ospreys through the binoculars as they went about the business of readying their nest on a buoy for the upcoming breeding season. I was intrigued by a pair of starlings darting in and out of the side of the massive structure of sticks and driftwood. Closer observation revealed the starlings were constructing their own nest in the loose framework of the older portion of the ernes' property. The pair of ospreys appeared oblivious to the constant coming and going of the smaller birds. I imagine, however, the starlings (as brash and obnoxious as they are capable of being) were careful to mind their manners so as not to impose or inconvenience the ernes. Subsequent observations revealed this commensal relationship (where one species benefits from association with another and the latter is neither harmed nor benefited) was proper and aboveboard as the starlings successfully reared their own brood to maturity in the downstairs area of the ospreys' home.

Ospreys are migratory to the extent that they move southward along the coast in winter, largely because the fish on which they prey head southward as the water becomes colder

The Return of the Erne

in the north. Ospreys on inland waters have to move southward, of course, when their northerly fishing grounds are frozen over.

The return of the erne to the Chesapeake has been a remarkable reversal of the general trend of many species of wildlife to fewer and fewer numbers—ending in extinction. I sincerely hope this bird has reestablished itself to such an extent that all future generations of nature lovers will have the opportunity and pleasure of observing it in its natural habitat and not as a stuffed skin behind glass in some museum collection.

11

The Big Adventure; Is It Worth It?

ON A CRISP, pristine November afternoon in 1978, I was on the bulkhead at Nanticoke Harbor watching the seemingly endless flights of ducks and geese migrating up and down the river. Though the birds were on the Dorchester side, my "7 × 50" binoculars enabled me to identify most of them. Black ducks, mallards, canvasbacks, and Canada geese were on the wing searching for an afternoon dining and resting area. I had never seen so many waterfowl before and expressed this to an elderly gentleman who had approached my observation site. "You should have been here fifty years ago," he exclaimed, and added, "In them days the sky was black with 'em, and you didn't need no decoys or fancy calling gadgets to kill all you wanted." He went on to relate how he used to guide weekend hunters from Baltimore, Washington, Philadelphia and New York, and it was not unusual for each man in the party to kill fifty to one hundred ducks every morning. These sportsmen would keep a few to show the folks back home, and the guide would retain a dozen or so for himself and for his neighbors, but the majority would be discarded in the river to decompose and float away. He recalled that sometime around 1940 it became more difficult to bag large numbers, so that he had to cheat a bit and use decoys. However, the decoys he used were nothing more than pieces of four-by-four lumber about eighteen inches in length. A screw eye and a bit of fishing line with a lead sinker attached served to anchor the bogus duck, and it

The Big Adventure

was routine procedure to set two to three hundred of these counterfeits in the shooting area. It is interesting to note that in recent times some duck hunting experts are recommending the use of large numbers of decoys instead of the usual two to three dozen. Apparently, ducks and geese sense safety in numbers and will pitch more readily in an area which seemingly has been found to be safe by numerous other flights.

The old man further explained there were bag limits in effect at the time, but no one really paid much attention to them. The rare game warden you encountered, he noted, seldom bothered a hunting party unless you got really "piggish." A hunter was usually all right as long as he did not try to dispose of his kill on the commercial market. This would definitely get him in trouble with the law. Every now and then a market hunter would be arrested and fined, but this did little to deter the unsavory practice. Finally, laws were enacted making it illegal to purchase wild ducks and geese as well as to serve them in eating establishments. These laws, more than anything else, almost caused the cessation of market hunting. But, even today, one hears rumors of "marsh rats" who supplement their incomes by the clandestine shooting of ducks and geese. Such an individual supposedly lives on the edge of the marsh near Tyaskin and employs one of the legendary "big guns" in his illicit work, but I doubt if there is any truth in this story.

Because the rules and regulations (both state and federal) concerning the hunting of waterfowl are so numerous and complex today, many hunters throw up their hands and choose to forego the pleasures to be obtained from the biggest adventure of all—the hunting of ducks and geese. Practically, it has come to the point where a lawyer is necessary to interpret the multitude of confusing laws strictly enforced by both state and federal authorities. In Maryland and several other states the bag limit is defined by a point value assigned to each species. Sometimes this value is quite different for the two sexes within the same type. For example, mallard hens carry a much higher point value than mallard drakes. The hunter is permitted to amass a given number of points (usually 100) in one day and cannot exceed another total for all the birds in his possession. For the last several years it has been absolutely forbidden to shoot some types like canvasbacks and redheads. Therefore, it is essential for the nimrod to be able to identify the different species as well as the sex. These sorts of rules have led to wanton waste rather than conservation. You may be sure a hunter is not going to be caught with a canvasback in his possession, or legal ducks which boost him over the point maximum. When a can (the common nickname for the canvasback) or redhead is accidentally gunned down, it will obviously be disposed of in short order. Likewise, if a duck is taken that puts the gunner over the point limit, it is very likely the bird will be hidden from the prying eyes of the law and not be counted. The hunter will then try to fill his quota with ducks of a lesser value.

For a number of years it was illegal in certain parts of Maryland (mainly the Eastern Shore) to use lead shot in 12-gauge shotguns over open water. Lead shot was permissible, however, in 10-, 16-, and 20-gauge guns. The rationale behind this ban on lead shot was that dabbling ducks (ducks that feed in shallow water and take in food and detritus from the bottom) gobbled up the spent lead pellets and were eventually poisoned. If you wanted to use the popular 12-gauge, it was necessary to employ shells containing steel shot.

The steel versus lead shot controversy raged in Maryland for several years with neither side able to prove its point. On one hand, the advocates of steel shot insisted the lead was decimating the waterfowl populations; on the other hand, the opponents of steel argued that many birds were lost to the game bag by the insufficient killing power of the steel (steel shot will frequently pass through the duck's body without distorting in shape whereas lead pellets will mushroom on entering the tissues and cause massive damage). The antisteelers also insisted, with some ballistic justification, that steel shot would ruin the barrel of a shotgun. One hunter along the Nanticoke told me he ruined an expensive dental bridge by biting down on a steel pellet left in the carcass of a duck upon which he was dining. Also, if steel is so superior to lead, the

Canvasback ducks over the marsh

advocates of lead wondered, why should steel be used only in the 12-gauge gun? The 10-gauge is actually a much better duck gun. It was difficult for many on the Eastern Shore to understand why the lead shot ban applied only to certain counties on the Shore and not to the southern and western portions of the state.

For the time being the controversy has been settled by the passage of a law sponsored by State Senator Frederick Malkus of Cambridge in Dorchester County that permits the use of lead shot in all gauge guns in any part of the state. Hunters on the Eastern Shore breathed easier during the 1980-81 hunting season, because in the years immediately preceding the "Malkus Law," a ducker was liable for arrest and a heavy fine if he was caught with so much as a 12-gauge shell with lead in his pocket and had a 12-gauge gun in his possession.

As an individual interested in biology, wildlife, conservation, and duck hunting I carefully researched the investigations—investigations which resulted in the conclusion that lead shot would, if its use continued, lead to highly deleterious effects on waterfowl populations. In my opinion, none of the research concerned with the effects of lead shot on ducks and geese was conclusive.

I traveled to the hallowed legislative chambers of Annapolis one cold day in January of 1979 to hear expert witnesses on both sides debate the issue, and I came away just as undecided about the matter as I had been before. Certainly, there seems to be no doubt that lead ingested by a dabbling duck or goose over an extended period of time can build up to a lethal concentration in the bird's body, but the studies and subsequent testimony fail dismally to prove that this buildup routinely occurs in the natural environment. Also, the argument of the antisteelers that many birds hit with steel shot are not killed cleanly and limp away to expire in an irretrievable place was not demonstrated to my satisfaction. One expert from the State Department of Natural Resources stated that steel would unconditionally kill as effectively as lead, and that the main cause of crippled birds was that the gunners were such poor

The Big Adventure

marksmen. This thesis did not sit particularly well with a number of veteran hunters present at the hearing. Many of these men take great pride in the fact they routinely drop nine birds for every ten shells fired.

The Malkus Law is a good one for our present state of knowledge. All too often regulations concerning hunting and fishing have been adopted not based on solid scientific data. Frequently, these ill-advised rules backfire and produce effects quite different from what was originally intended. Therefore, controlled experiments by competent investigators are urgently needed in order to arrive at the truth.

When thinking about the legal bickering and the mumbo jumbo concerning duck hunting, I recall an expedition I took in 1978 with Bob Mitchell of Mt. Vernon. Mitchell and I had scouted the area near Newfoundland Point for several days and noted many ducks active in a small, semisheltered cove. We arrived on the scene just before daybreak on a cold December day, placed the decoys, and erected a portable blind just as the first rays of the new day's sun broke through the low cloud cover. It was a day that held great promise of success as everything seemed to be right for a productive and satisfying hunt. We had taken every precaution to make sure we were legal hunters. We had the necessary state hunting license, special state duck stamp, stamp for hunting on public land, federal duck stamp, a booklet giving the point values of the various species, and—just to make sure—a Peterson guide to the birds of the eastern United States. We had no 12-gauge shells containing lead shot in our possession. We each brought two guns—Mitchell was armed with a 20-gauge pump and a 12-gauge automatic, and I had an automatic 16-gauge and a 12-gauge goose gun. We made sure the guns were plugged in such a manner that they were incapable of holding more than three shells at one time. (This is a federal law applying to the hunting of any migratory species of bird.) We had ammunition containing lead pellets for the 16- and 20-gauge guns and shells with steel shot for the 12-gauge guns. And, of course, our numerous stamps and documents were pinned to the backs of our hunting coats as required by law. What an arsenal! You might think we were taking up a position on the Nanticoke to defend it against a foreign invader rather than just trying to get a legal shot at a duck or two.

After camouflaging the boat and brushing up the blind, we settled in to await some action. We did not have long to wait. Just as I was taking my first guarded sip of scalding hot coffee, Bob nudged me and nodded in the direction of Tyaskin Harbor. A group of goldeneyes were heading our way, and in the stillness of the morning we could hear distinctly the high-pitched "wh-wh-wh-wh" of their whistling wings. I reached for the automatic sixteen, and Bob flipped the safety off his twelve. The flock headed straight for the decoys and gave every indication of pitching among them. At about seventy yards out, however, they banked suddenly to the right and headed south, down the river. We thought, "Oh well, that's the way it goes sometimes." Something had alerted the whistlers to our presence. It could have been a ray of sunlight reflected off a gun barrel, an unconscious movement on our part, or an unnatural bob of one of the decoys.

We settled back in the blind to await further developments and vowed to make sure we remained totally concealed until the next incoming flight had decided irreversibly to come within range. Again, the wait was not long. Within five minutes I spotted a flight of either black ducks or mallards zooming in on us just above the water's surface. Bob whipped out his calling device and encouraged them toward us with an expert rendition of the greeting call. The ducks reacted in a positive manner, and I felt sure we would have a chance with this group. With the adrenaline being released in our bodies in gargantuan amounts, we watched those ducks make a beeline for our decoys, but the same disaster occurred again. At about fifty yards (still out of effective shooting range) they, too, suddenly veered off to the right and hightailed it for parts unknown. What a letdown. Profoundly dejected we were at a loss to explain what had happened. During the next hour or so the same series of events transpired two additional times, and we were ready to pack up and call it a day. Something was

The Big Adventure

wrong, and we were frustrated by our inability to determine what it was. After all, we should at least be as clever as a flock of birds! I even asked Bob if he had forgotten to use his deodorant that morning, but the question evoked no laughing response.

We were in the process of unloading and casing the guns when a lone drake mallard seemingly came out of the blue, zipping along the shoreline. I saw him coming and reached for the goose gun which was still loaded with steel shot. This weapon has a forty-three inch barrel with a tightly closed full choke and will maintain an effective pattern of shot for up to sixty yards. Because of its extended range, many people refer to the gun as the Long Tom. As the duck approached he was on a line out from the blind about twenty yards. Approaching the decoys, he also banked sharply away and turned on his afterburners in a direction away from us. He made his move too late, however, as he was only about forty yards away when I squeezed the trigger on the Long Tom. He dropped to the water like a rock, and I felt a great sense of satisfaction in that we would not come home completely empty-handed that day. As I paddled out to retrieve the bird I could hear the muffled thump, thump, thump of guns being fired up and down the river, presumably at ducks, and I could not help but wonder why others were getting an abundance of shots, and we had had only one. Just as I was fishing the duck out of the water I was startled by the roar of an outboard motor being coaxed to life in a small secluded cove some one hundred fifty yards away—suddenly, I had the answer to the question that had plagued us all morning!

An olive drab boat bore down on us. For a split second I thought the man at the controls was intent on ramming us. He pulled up short by reversing the engine with a great backwash of water and flashed a wallet containing a badge that identified him as a Federal Wildlife Conservation Officer. He requested permission to check us out. Obviously we were spotted earlier (probably as we set up the blind), and he had idled his boat into a position in the cove where he was out of our line of vision. He had been there in hiding for the entire morning and certainly was responsible for spooking the ducks as they headed in our direction.

The officer was overly courteous and businesslike as he went about his task of trying to ferret out some violation of the law on our part. Every inch of the boat was inspected for illegal birds or shells, as was the inside and immediate outside area of the blind. He stooped once to pick up a spent 12-gauge shell casing which had contained lead shot, but he quickly discarded it when a sniff indicated it had been fired some time ago. The guns were inspected for the mandatory three-shot plug, and two boxes of ammunition that had the words "Steel Shot" printed on the outside were opened and each of the shells scrutinized. While going over our papers it was pointed out that Mitchell had neglected to affix his signature across the face of the federal stamp pasted on the back of his license. The officer allowed, however, that if Bob would sign the stamp then and there, he would not pursue the matter.

I asked him why he had not checked us out earlier and called to his attention the fact his prolonged presence near our blind had deprived us of several opportunities to shoot. He laughed at this, replying with remarks to the effect he had been just waiting to see if we would shoot an illegal canvasback or redhead. He stated he had information indicating cans were being routinely shot all along the Nanticoke, and he intended to put a stop to it.

With that he wished us "good shooting" and was back in his boat, apparently out of our hair and on his way. However, he did not leave the area immediately. With one other thing to investigate, he took a few minutes to drag a fine-meshed scoop along the bottom of the river where our decoys were placed. He was looking for corn or some other bait we could have been using illegally to attract the birds. Fortunately for us he found none. During the dredging operation, however, we were nervous as we knew the area the day before had been the hunting site of two men from Tyaskin who were known to use corn routinely as an illegal bait when they shot over decoys. I seriously doubt if we would have been able to convince the warden that any bait he found had not been placed there by us, and I

The Big Adventure

feel sure he would have arrested us on the spot if he had found any grain in the water. He would have transported us to Baltimore for arraignment, confiscated our guns and licenses, and required us to pay stiff fines.

By this time the sun was well up into the early winter sky, and our decoys were reflecting the solar rays like mirrors. We knew the hunting was finished for that day. On the boat launching ramp back at Wetipquin we noticed a trash can full of feathers and entrails that had formerly made up the anatomies of several canvasbacks. The remains were quite fresh and had been deposited in the can not more than one or two hours previously.

Some hunters will risk arrest and shoot canvasback, or redheaded mallard as some call it, because it is regarded by many as the "King of the Waterfowl"—so-called because of the savory flavor of its flesh. This delectable taste is due to the eel grass (also known as wild celery) which the canvasback ingests. In recent years there has been a dramatic decline in eel grass over the entire Chesapeake Bay system, and biologists have been unable to determine just why this has occurred. In any event, the reduction in eel grass is most certainly correlated with the decimation of canvasback populations.

It is said the famous canvasbacks of the Chesapeake Bay had a great deal to do with the designation of Baltimore as the "Gastronomic Capital" of America. Too much of a good thing, however, can lead to a lack of appreciation of its finer qualities. For example, it is reported that an old contract between two slave owners in Tidewater Virginia stipulated that when one owner hired slaves from the other he was not to feed them canvasback ducks more than five times a week. I would hazard a guess the slaves' diet was supplemented with diamondback terrapin on the days when duck was forbidden. As mentioned earlier, too much terrapin in the rations was also known to cause unrest among the slaves.

Those few hunters who deliberately down illegal species must be on the alert for specially equipped aircraft utilized by enforcement officials. The location of a suspicious hunting party will be radioed to a nearby patrol boat to check on an operation of which the officer on the water was unaware. I have had these spotter planes buzz my blind no more than fifty feet above the ground, and, of course, this sort of shenanigan does little to improve hunting conditions.

As Bob and I were hauling the boat out of the water that morning (managing at the same time to take a considerable amount of ice cold creek water into our hip boots) he turned to me and asked, "Is it worth it?" In a flash I remembered the tremendous thrill and exaltation at the sight of that first flight of whistlers coming in, and all the frustrations and bitterness of the day vanished. I replied, "You bet it was!"

As noted, many people in our area of the Nanticoke and elsewhere are saying the cost of shells, decoys, licenses, gasoline, and other necessary paraphernalia is so high these days they cannot afford to go ducking any longer. Add to this monetary burden what many consider to be harassment by enforcement officials, and one can readily see why many old-time hunters (as well as the beginners) have decided to call it quits.

But, as for myself, I guess I will never give up hunting as long as I have the strength and resources to do it. And, a considerable amount of strength and intestinal fortitude is necessary to venture out on the river in quest of waterfowl during the foulest times of the year when it is legal to hunt. I can recall several times breaking ice in the shallow water to maneuver the boat into and out of a shooting area, then sitting hunched down in a crowded blind in freezing weather trying to prevent frostbite and watching for ducks at the same time. Many outdoorsmen feel a pint or so of hard liquor is as essential to a duck hunting trip as the shells, guns, and decoys. Sometimes the fiery antifreeze seems to improve the hunting as well as giving the illusion of warming the body.

Wendy Watson of Berlin, Maryland tells the story of a duck hunting party on the Nanticoke a few years ago where the ardent spirits flowed more freely than the birds flew. One hunter who had been constantly nipping at his pint-sized hip flask suddenly reared up in the blind and fired at a bird which was so high it was "scraping the sky." As will happen once in a

The Big Adventure

thousand times with a "sky-busting" shot, the duck fell to the cheers of admiration from the others in the party. The sharpshooter would not accept any plaudits, however, and is reported to have replied, "Hell, that wasn't such a good shot. I should have gotten four or five out of that flock!" It has long been my contention though that anyone who permits the consumption of alcohol in or around a duck blind has certainly taken leave of his senses.

Duck hunting along the Nanticoke is not without personal risk from the elements. The river is unforgiving of human carelessness, and one must learn to respect it at all times. One day in December 1979 I was sitting on the bulkhead in front of the cottage watching numerous flocks of birds working about a mile away along the opposite shoreline. I suddenly had the overpowering impulse to take the boat down to Wetipquin Landing, put it over, and head for the other side either to take some photographs or perhaps bag a fat mallard for supper. Without realizing it was 3:00 P.M. and only a little more than two hours of daylight remained, I hastily grabbed a camera, shotgun, and a few shells and sped off to the launching ramp. Quickly I got the boat into the water and headed out of Wetipquin Creek for Dorchester County. As I approached mid-river a tugboat pushing a barge was making its way downstream from either Seaford or Vienna through the deep channel that runs close to the western shore. This intrusion on their bailiwick disturbed the hundreds of feeding birds (the blowing of the tug's horn by the captain to signal his presence to me didn't help matters either), and they took to the wing and scattered.

I went ashore at Captain John's Beach and waited for at least some members of the flocks to return. When the birds failed to materialize after about half an hour, I glanced at my watch and noted with a considerable degree of apprehension that it was 4:30 P.M. No problem, I thought, as it was only a fifteen to twenty minute run back to Wetipquin, and I should have the boat out of the water and be back at the cottage about 5:30 P.M. However, this timetable was not to be, as about halfway across the river the outboard motor sputtered and coughed a few times before quitting entirely. As luck would have it there were no other boats visible which might come to my assistance, and the motor absolutely refused to be restarted (I later learned a reed valve between the cylinders had broken). I was at the mercy of the wind and the tide, and about the only thing I could do was attempt to guide the boat with a paddle as it was blown, ever so slowly it seemed, toward a desolate section of beach north of the entrance to Tyaskin Harbor.

The sun had already disappeared below the horizon when the boat was finally blown onto the shore, and I knew it would be pitch dark in a matter of fifteen or twenty minutes. Though the day had been fairly warm, the night breeze felt like frigid daggers as it pierced my wet clothing, and I was beginning to shiver all over. Taking stock of the situation I realized what a fool I had been to set out alone that afternoon. I had not even had enough sense to tell a neighbor of my plans, and in my haste to get going I had not bothered to take along any emergency gear—not even a flashlight or signal flare. Such actions are inexcusable.

Where the boat drifted ashore, the marsh abuts the beach but without a light it would have been impossible to walk out either along the beach or through the marsh. By this time as the tide was full, the beach itself was almost nonexistent, and to try to hike out through the marsh would surely result in at least one broken leg from stepping in a muskrat hole. I did have an old piece of tarpaulin in the boat, two cigarettes, and an inexpensive butane lighter that was almost out of gas. By now it was 5:30 P. M. and totally dark. I resigned myself to spending the night on the edge of the marsh and could only hope I would be able to ward off hypothermia by somehow getting a fire going with cord grass, wet driftwood, and gasoline from the outboard's tank. To compound my problems further the lighter spewed out its last molecule of butane and would not produce a flame. I thought I could perhaps use the spark from the flint and steel mechanism of the lighter to ignite gasoline-soaked grass, but I would have to be very careful and not have the gas blow up in my face. Just as I was about to attempt to

The Big Adventure

ignite a fire I heard the faint sound of an outboard motor somewhere out on the river. Oh, what I would have given for a flashlight or emergency flare about that time. The boat seemed to be coming in my direction, and so I yelled for help at the top of my lungs. There were no indications from the boat that my shouts had been heard. I felt in my pocket and found the six shotgun shells I had stuffed in there in my initial haste to get going earlier in the afternoon. I quickly loaded the gun with fingers that were numb with the cold and fired the three-shot salvo I remembered was supposed to be an international signal of distress. Still the boat hummed along giving no indication my desperate plea for help had been heard. I was just about ready to fire the last three shots—I could not let the boat pass me by—when the motor spit a couple of times and died. I could hear faint curses coming from the occupants of the vessel, and I new my saviors were not far away. I fired the last three rounds, and almost immediately a high-powered spotlight swept the length of the beach. I was waving my arms frantically and shouting when the beam focused on me. The boat immediately proceeded to the beach, and the two occupants pulled me over the gunwale. They explained the motor had quit because one of the gas tanks had run dry. The men had spent the afternoon hunting in the vicinity of Penknife Point and had not checked the level in the tank before they headed in. The switch of the gas line from the empty tank to a full one was made quickly and the motor was immediately operable. They had not heard my distress signals while the motor was operating and had no idea I was stranded nearby. They graciously secured a tow line to my disabled craft, and we headed back to Wetipquin landing two or three miles away.

After expressing my everlasting gratitude to the men (they simply replied that I might do the same thing for them sometime in the future) I secured the boat to the pier and headed back to the cottage with the car heater going full blast. I made it home by 6:45 P. M., had some hot coffee, and jumped into bed under three blankets. One thing I learned from this big misadventure was never to take on the river without making adequate preparations. It was also very satisfying to realize that the "Code of the River," which says one is morally bound to come to the aid of another in distress, is taken seriously by those who venture forth upon the water.

So, in spite of all of these mishaps, I still firmly believe the "Big Adventure" is worth it. There has never been a ducking trip that I did not learn something new about the nature of the species or some other facet of natural history. To my way of thinking there is really a lot more to duck hunting than killing ducks. I have always been thrilled and emotionally moved when I see a flock of waterfowl pass overhead. Their unique flight patterns are singular studies in natural symmetry and beauty, no two patterns exactly alike. The grace and artistic agility the birds exhibit when coming in for a landing or running lightly across the water's surface in preparation for taking to the air are natural vignettes that have never been captured fully on canvas or film. These flight patterns are phenomena of nature one has to view firsthand in order to appreciate.

I cannot begin to estimate the number of hours I have spent "talking" with ducks. I have a weakness for duck calls, and every time I see a new one in a store or advertised in an outdoors magazine I simply must have it. Although duck calls all sound pretty much the same to me the way I blow them, after acquiring a new call I can hardly wait to get in position along the river and try it out. Just to be able to speak part of the duck's language, draw him in with a greeting or feeding call, and then watch him reverse his line of flight when you put a good comeback on him is a tremendous thrill.

One of the finest duck clubs along the Nanticoke is located just inside the mouth of Jack's Creek (there is even a television antenna protruding from the roof of one of the buildings), and a longtime member of this organization is a wealthy Salisburian. This individual, an odd duck (to make a pun) in that he does not own a gun, frequents the club as often as its shooting members do. He maintains his membership solely for the joy of calling ducks, and is, as one might suspect, an expert in waterfowl linguistics. A story is told of a game warden who checked up on him early one morning as he sat in his

The Big Adventure

blind conversing with his feathered friends. The warden would not believe this gentleman was there simply to observe and call ducks. He was sure the man had to be up to some mischief or was crazy. After a conversation revealed the gentleman was completely sane and since no illegal weapon could be found, the warden went on his way mumbling to himself with the quack-quack-quack of the greeting call and tickeeet-tickeeet-tickeeet of the feeding sound undoubtedly ringing in his ears.

The waterfowl along the Nanticoke and in the other parts of the Chesapeake may not darken the sky with their sheer numbers as they did fifty to seventy-five years ago, but they seem to be holding their own at the present time from a numerical standpoint. Somehow, I feel confident there will be enough waterfowl around fifty years from now to provide sport and enjoyment for the members of that generation. The numbers of some species, like the canvasback, redhead, and wood duck, are at present somewhat depleted, but other species like the noble Canada goose and bluebill duck have, according to some census figures, increased in numbers during the last thirty to forty years. Species have their periodic ups and downs in regard to population numbers, and these fluctuations are inevitable in a changing environment. Most species, however, have built into their genetic makeup the latent variability that enables them to adapt to new and challenging environmental situations—thus insuring the perpetuation of their kind.

If the conservation laws are enacted with care and enforced in the proper manner, I believe we can rest assured the Big Adventure will be available for a long time to come to all of those who care to seek it.

Appendix

Scientific Names of Animals Included in the Text Listed in Phylogenetic Order

Phylum: *Coelenterata*
    Lion's mane jellyfish   *Rhopilema verrilli*
    Moon jelly   *Aurelia aurita*
    Portuguese man-of-war   *Physalia physalis*
    Red winter jellyfish   *Cyanea capillata*
    Sea nettle   *Chrysaora quinquecirra*
    Sea wasp   *Carybdea alata*

Phylum: *Ctenophora*
    Comb jelly (sea walnut, comb bearer)   *Pleurobrachia brunnea*
    Jug-stopper   *Beroe cucumis*

Phylum: *Annelida*
    Bloodworm   *Glycera dibranchiata*

Phylum: *Mollusca*
    Fresh water clam   *Unio* sp.
    Oyster   *Crassostrea virginica*
    Sea (hard) clam   *Mercenaria mercenaria*
    Scallop   *Pectin* sp.

Phylum: *Arthropoda*
  Insecta
    Blue-tail fly   *Tabanus atratus*
    Deerfly   *Chrysops* sp.

Greenhead fly	*Tabanus nigrovittatus*
Horsefly	*Tabanus* sp.
Housefly	*Musca domesticata*
No-see-um (punky, sand fly)	*Culicoides* sp.
Salt-marsh mosquito	*Aedes sollicitans*
	Ae. taeniorhynchus
Stable fly	*Stomoxys calcitrans*

Merostomata
   Horseshoe crab — *Limulus polyphemus*

Arachnida
Chigger	*Trombicula* sp.
Tick	*Amblyomma* sp.
	Dermacentor sp.

Crustacea
Blue crab	*Callinectes sapidus*
Cyclops	*Oithona brevicornis*
Grass shrimp	*Palaemonetes vulgaris*
Longhorn	*Acartia clausi*
	A. tonsa
	Eurytemora affinis
Water flea	*Podon polyphemoides*

Phylum: *Echinodermata*
Sea cucumber	*Thyone* sp.
Sea star (starfish)	*Asterias* sp.

Phylum: *Chordata* (Subphylum: *Vertebrata*)

Chondrichthyes
Butterfly ray	*Gymnura* sp.
Cownose ray	*Rhinoptera bonasus*
Eagle ray	*Aetobatis* sp.

Osteichthyes
Alewife	*Alosa pseudoharengus*
Blueback herring	*A. aestivalis*
Bluefish	*Pomatomus saltatrix*
Carp	*Cyprinus carpio*
Catfish	
Channel	*Ictalurus punctatus*
White (Potomac)	*I. catus*
Garfish	
Alligator	*Lepisosteiformes spatula*
Longnose	*L. osseus*

Appendix

Shortnose	*L. platostomus*
Spotted	*L. productus*
Grayling	*Thymallus signifer*
Hardhead (croaker)	*Micropogon undulatus*
Menhaden (bunker)	*Brevoortia tyrannus*
Norfolk spot (spot)	*Leiostomus xanthurus*
Redhorse	*Moxostoma* sp.
Rockfish (striped bass, striper)	*Morone saxatilis*
Sea robin	*Prionotus strigatus*
Shad	*Alosa sapidissima*
Sturgeon	*Acipenser* sp.
Toadfish (oyster toad)	*Opsanus tau*
Trout (weakfish)	*Cynoscion regalis*
White perch (black perch)	*Morone americana*

Reptilia
Alligator snapping turtle	*Macroclemys temmincki*
Black snake	*Coluber constrictor*
Copperhead	*Agkistrodon contortrix*
Cottonmouth	*A. piscivorus*
Diamondback terrapin	*Malaclemmys terrapin*
Hognose snake (hissing adder)	*Heterodon platyrhinos*
Northern water snake	*Nerodia sipedon*
Snapping turtle	*Chelydra serpentia*

Aves
Bald eagle	*Haliaeetus leucocephalus*
Black duck	*Anas rubripes*
Bluebill (scaup)	*Aythya affinis*
Buzzard	
Black	*Coragyps atratus*
Turkey	*Cathartes aura*
Canada goose	*Branta canadensis*
Canvasback duck	*Aythya valisineria*
Goldeneye duck (whistler)	*Bucephala clangula*
Great blue heron	*Ardea herodias*
Mallard duck	*Anas platyrhynchos*
Osprey (erne, sea eagle)	*Pandion haliaetus*
Purple martin	*Progne subis*
Redhead duck	*Aythya americana*
Starling	*Sturnus vulgaris*
Swan	*Olor columbianus*
Wood duck	*Aix sponsa*

Mammalia
- Fox — *Urocyon haliaetus*
- Muskrat (swamp rabbit) — *Ondatra zibethicus*
- Opossum — *Didelphis marsupialis*
- Raccoon — *Procyon lotor*
- Skunk — *Mephitis mephitis*
- Whitetailed deer — *Odocoileus virginianus*

Index

A
Abatement programs, 25
Aerie, 138
Alaskan king crab, 30
Alewives, 79
Allen, Jessie, 106, 110-11
Alligator garfish. *See* Garfish
Alligator snapping turtle, 129
Anadromous species, 82, 84
Arachnid theory, 40
Archaeopteryx, 35
Arsenic, 115
Arthropods, general, 31
Assateague Island, 34
Atlantic Ocean, 18, 24
Autotrophs, 10

B
Bait
  for crabs, 80, 82
  to attract waterfowl, 152
Bald eagle. *See* Eagle
Baltimore, Maryland, 4, 63, 153
Barge (river), 6, 155
Barrow, Alaska, 46
Bay bridges, 35
Beaufort, North Carolina, 123
Bedloe, Ed, 46
Benzocaine, 24
"Big Gun," 145
Biogenetic Law, 39
Biological control, 59
Bioluminescence, 28
Bivalve Harbor, 20, 22, 65
Bivalve, Maryland, 6, 20, 49, 50
Blackbeard the pirate, 4
Black duck, 144, 150
Black perch, 80
Blacksnake, 134
Bloodworms, 94, 110
Bloodsworth, Jessie, 21
Blueback herring
  fishing for, 78, 84-86
  identification of, 79
  life cycle and migration of, 80, 82-84
  synonyms for, 79, 84
  uses of, 79, 82
Bluebill, 159
Bluebottle, *See* Portuguese man-of-war
Blue crab, 36
Bluefish, 68, 80
Blue racer, 134
Blue-tail fly, 62
Brady, Diamond Jim, 120
Brazos River, Texas, 71, 75, 90
Butterfly ray (sand skate), 102
Buzzard, 141

C
Canada geese, 144, 145, 148, 159
Cannon, Patty, 4
Cannon's ferry, 4

Canvasback, 144, 152, 153, 159
Cape May, New Jersey, 35
Captain John's Beach, 155
Carp, 69
Catfish
   envenomization by, 95-97
   fishing for, 89-90
   preparation of as food, 93, 94-95
   treatment of stings from, 95-96
   types and identification of, 88, 89, 92, 93
Catts, E. P., 57
Cedar Hill Park, 20, 22
Cell layers, 17
Chesapeake Bay, 13, 18, 43, 70, 82, 88, 94, 103, 136, 153
Chesapeake perch, 92
Chiggers, 62
Chitin, 36, 38
Chlorinated hydrocarbons, 137. *See also*, DDT
Choptank River, 35, 112
Chrysops. See Deerfly; Greenhead fly
Citronella, 49, 50, 51
Clams, 37
"Code of the River," 158
Colloblasts, 27, 28
Colville River, Alaska, 53
Comb bearer, 27
Comb jelly, 27
Commensal relationship, 142
Comparative anatomy, 40
Copepods, 9, 11, 13, 14
Copperhead snake, 130, 133, 134, 135
Cottonmouth snake, 130
Cownose ray
   distribution of, 103
   feeding habits of, 103
   fishing for, 103-4
   use of as food, 104
"Crabphobia," 36
Crab pots, 18, 36, 80, 104, 119, 126
Crabs, 20, 36
Croaker. *See* Hardhead croaker

D

Daphnia, 11
DDT, 137, 140

Decoys, 144
Deep Creek, 3
Deer. *See* White-tailed deer
Deerfly, 62
Delaware Bay, 18, 35
Delaware River, 94
Delmarva Peninsula, 3, 18, 35
Devil crab, 35
Diamondback terrapin
   distribution, 123
   preparation of as food, 120, 126-27
   types and identification of, 21, 119, 123, 124, 125
   uses, 120, 122, 126
Diatoms, 9
Diethyl toluamide (repellent), 54
Dinoflagellates, 9, 12
Dissolved oxygen, 12
Distress signal, 157
DNA (deoxyribonucleic acid), 43, 113
Dog fly. *See* Stable fly
Dredging, 8, 37
Duck calls and calling, 158-59
Duck clubs, 158

E

Eagle, bald, 11, 136-37, 140-41
Eagle ray, 103
Ecology, 8, 26, 51
Eelgrass (wild celery), 153
Eels, 37, 133
Ellis Wharf, 4
Emperor's Landing, 4
Endangered Species List, 138
Endotoxins, 38
Erne (ern). *See* Osprey
Ernst, Manfrid, 60
Ethyl hexanediol (repellent), 55
Evolution, 16, 39, 61, 127

F

Fish fry, 11
Fish hawk. *See* Osprey
Fish kills, 11-12
Flagellates, 9
Flatworms, 17
Fly season, 57, 61

Index

Food chain, 10-11, 26
Fort Worth, Texas, 69
Fox, *xvi*
Freon gas, 56
Fyke nets, 126

G
Game fish, 75
Ganoid scales, 71
Garfish
  distribution, 65, 68
  eggs, 70, 88
  evolution, 71
  fishing for, 64, 65, 67, 72
  preparation of as food, 73-74
  synonyms, 68
  types and identification of, 64, 68, 69, 70, 73
  uses, 70, 77, 88
Germany, 35
Gill nets, 64, 80, 106, 107, 111
Glochidia (clam larvae), 77
Glycolysis, 125
Goldeneye duck (whistler), 150
Grass shrimp, 108, 114
Gravelly Branch, 3
Grayling, 53
Great blue heron, 11
Great Shoals Light, 103
Greenhead fly
  bite of, 54
  control of, 55-56, 57-58
  life cycle of, 56-57
  types and identification of, 42, 56
Greeting call, 150
Griffin, Lou, 26, 92
Griggsville, Illinois, 59
Gulf of Mexico, 11, 24, 35
Gum Branch, 3
"Gut Puller," 108

H
Hansens, Elton J., 57
Hardhead (croaker), 65, 68, 76, 100
Hatcrown Point, 65
Herring family, 79
Herring Run, 3
Heterotrophs, 10
Hissing adder. *See* Hognose snake

Hognose snake, 134-35
Holloway, Luther F., 74
Horn Point Biological Lab, 35, 115
Horsefly, 54
Horseshoe crab (*Limulus*)
  evolution of, 31
  life cycle of, 30, 32
  uses of, 35-36, 37-38
Housefly, 60
Hudson River, New York, 115
Hurley Drain, 3
Hurrican Agnes, 84
Hush puppies, 89, 90
Hypothermia, 156

I
Indian Ocean, 18
Insecticidal applicators, 48
Insects, general, 42-44

J
Jack's Creek, 48, 132, 158
Jackson Harbor, 65
Japan, 18
Jellyfish, 17, 20, 25
Jones, Tom, 35
Jug-stoppers, 27-29

K
Kentucky Lake, 89
Krantz, George, 115

L
Lepisosteus osseus. *See* Garfish
Lewis Landing, 4
Limulus. *See* Horseshoe crab
Lion's mane jellyfish, 23
Longhorns, (copepods), 14
Long Point, 3, 6
Long Tom (shotgun), 151

M
Mahogany tide, 12
Malkus, Frederick (Senator), 148
Malkus Law, 148-49
Mallard duck, 144, 150, 151
Manokin River, 112
Marsh grass, 124
Marshyhope Creek, 5

Menhaden, 79, 104
Mentasti, Ulysses, 100
Maryland Department of Natural Resources, 117
Maryland Institute of Marine Science, 22
Marine Police, 20, 111
Meat tenderizer, 22
Mexico, 123
Mississippi Valley, 70
Mitchell, Robert L. (Bob), 100, 149
Mites, 31
Mollusks, 12
Moon jelly, 23
Morone saxatilis. See Rockfish
Mosquito abatement agency, 56
Mosquitoes
  distribution of, 43, 44, 47
  importance of, 52, 53
  protection from bites of, 47-51
  types and identification of, 42, 44
Mount Vernon, Maryland, 100, 149
Muskrat, *xvi*

N

Nanticoke Harbor, 6, 144
Nanticoke Indians, 3, 47
Nanticoke, Maryland, 6, 49
Nanticoke River, 3, 5, 8, 18, 60, 68, 82, 93, 103, 115, 136
Natural history, 8
Nematocysts, 17, 20, 24
Nettles. *See* Sea nettle
New England, 48, 123
Newfoundland Gut, 133
Newfoundland Point, 149
Nightcrawlers, 94
Norfolk spot, 65, 68
Norfolk, Virginia, 130
North Carolina, 130
Northern water snake, 130, 132, 133
No-see-ums, 62

O

Ocean City, Maryland, 37
Oil producing gland, 141
Opossum, 47
Osprey
  eggs of, 137, 141-42
  habits of, 138, 140, 141, 142
  types and identification of, 136, 141, 142
Oyster spat, 11, 29

P

Padre Island, Texas, 24
Paleozoic era, 32
Pan crabs, 35
Parallel evolution, 61
Penknife Point, 35, 157
Peritoneum, 79
Philadelphia, Pennsylvania, 5, 63, 126, 128
Photosynthesis, 10, 12
Phytoplankton, 10-11, 13
Pit vipers, 132
Plankton, 9, 80, 116
Pocomoke River, 103
Poliomyelitis, 62
Pollutants, 116, 137
Polychlorinated biphenyls (PCBs), 5
Portuguese man-of-war, 24
Positive geotropism, 31
Potomac River, 112, 115
Priest (small club), 67
Public Landing, Maryland, 100
Punk, 49-50
Punkies, 63
Purple martin, 59

Q

Quantico Creek, 84

R

Raccoon, 11, 47
Radioactive tracers, 47
Ransbottom, Jack A., 74
Redhead duck, 146, 152, 159
Redhorse minnow, 71
Red tide, 11
Red winter jellyfish, 23
Rehoboth Beach, Delaware, 36
River turtles, 21
Robinson, Bill, 68
Rockfish
  conservation of, 107, 115-16
  distribution of, 110-11
  types and identification of, 106
"Rock fights," 113

Index

S

Salinity, 64, 65
Salisbury, Maryland, 63, 134
Salisbury State College, 35, 74, 100
Salmon, 85
Salt-marsh mosquito, 44, 46-47. *See also* Mosquitoes
Scallop (*Pectin*), 104
Scorpions, 31
Sea cucumber, 40
Sea eagle. *See* Osprey
Seaford, Delaware, 3, 5, 6
Sea nettle
 distribution, 18, 20
 sting and treatment for, 21, 22
Sea robin, 88
Sea star theory, 40
Sea walnut, 27
Sea wasp, 25
Selenium, 115
Shad, 79, 84, 86
Sharks, 25, 103
Sharptown, Maryland, 5-6
Shellfish, 12
Shot, lead vs. steel, 146-48
Shotguns, 146. *See also* Long Tom
Silent Spring, 59
Silverside minnow, 13
Skunk, *xii*
Smith, Captain John, 3
Smudge pots, 47
Snakebite and treatment of, 134
Snakefish, 69
Snakes, 130. *See also* individual types
Snapping turtle
 bite of, 128
 catching, 129
 distribution, 127
 habits, 127
 laws pertaining to, 129
 preparation of as food, 128, 129
Spiders, 31
Spring tide, 31
Stable fly
 control of, 61
 evolution, 42
 habits, 61
 identification of, 60-61
 medical importance of, 62

Starling, 142
Statocyst, 27
Stinging cells. *See* Nematocysts
Stingrays. *See also* Cownose ray; Eagle Ray
 stings and treatment of, 102
 types and identification of, 88, 101-2, 103
Striped bass. *See* Rockfish
Sturgeon, 70
Sussex County, Delaware, 3
Swan, *xv*
Swim bladder, 72, 101

T

Tangier Sound, 3, 6, 93, 136
Terrapin defined, 120. *See also* Diamondback terrapin
Terrapin dredge, 125
Thermodynamics, Laws of, 11
Ticks, 31, 63
Tidewater, Virginia, 153
Toadfish
 culinary qualities of, 100
 distribution, 88, 101
 habits, 98, 100-1
 venom apparatus of, 97-98
"Trashfish," 73
Trilobite, 39
Trout, freshwater, 53
Tugboats, 155
Turtles. *See* Diamondback terrapin; Snapping turtle
Tyaskin, Maryland, 6, 31, 49, 52, 145
Tyaskin Wharf (Harbor), 6, 68, 150, 156

U

Umiat, Alaska, 46, 53
U. S. Army Corps of Engineers, 8
U. S. Bureau of Fisheries, 123
U. S. Coast Guard, 5, 138, 140
U. S. Fish and Wildlife Service, 115, 138, 151
University of Delaware Institute of Marine Science, 104
University of Maryland, 115, 119
Urine, 22, 25, 50

V

Venom, 21, 22. *See also* under specific animal
Vienna, Maryland, 4, 6
Vitamin D, 116

W

Walnut Landing, 4, 5
Water fleas, 9, 11
Waterfowl
  conservation laws for, 145, 149
  flight patterns of, 158
Water snake. *See* Northern water snake
Watson, Wendy, 154
Weakfish. *See* Trout
Wetipquin Creek, 6, 49, 68, 78, 84, 153, 155, 157
Wheatley's Wharf, 4
Whistler duck. *See* Goldeneye
White perch, 29
White-tailed deer, *xv*
Wicomico County, Maryland, 20
Wicomic River, 100, 103, 112
Wood duck, 159
World War II, 50, 137

Z

Zooplankton, 10, 12-13